The A–Z of

Global Warming

THE A–Z OF GLOBAL WARMING

BY

SIMON ROSSER

Schmall World
Publishing

First published in 2008 by Schmall World Publishing
Fiveways House, 20 Duffryn Road, Cyncoed, Cardiff, CF23 6NP

www.a-zofglobalwarming.com

Distributed by Gardners Books
1 Whittle Drive, Eastbourne, East Sussex, BN23 6QH
Tel: +44 (0)1323 521555, Fax: +44 (0)1323 521666

British Library Cataloguing in Publication Data
A catalogue record for this book is available from the British Library.

ISBN 978-0-9558092-0-0

Typeset by Amolibros, Milverton, Somerset
This book production has been managed by Amolibros
Printed and bound by T J International Ltd, Padstow, Cornwall, UK

To my parents

CONTENTS

ACKNOWLEDGEMENTS

In alphabetical order, the author would like to thank the following people who helped in the creation of this book:

Andrew Marshall (GM Motors), Sir Crispin Tickell GCMG KCVO, Claude Duval, Emily Duval, Harold N Burdett (The Population Institute), John Rosser, Jan Rosser, Jane Tatam (Amolibros), John Brien, Lucy John, Mary Duval, Peter Cowlam, Rhett Butler (Mongabay) Russ Billington (illustrations), Renate Christ (IPCC Secretariat), Susan Rosser, Simon John and Zuzana Ivancikova.

And the following organisations/publications:

International Panel on Climate Change (IPCC), Mongabay, National Aeronautics and Space Administration (NASA), National Oceanic and Atmospheric Association (NOAA), National Snow and Ice Data Center (NSIDC), The Stern Review on the Economics of Climate Change, The Sacramento Bee Newspaper, California, and The World Wildlife Fund (WWF).

"…a useful simplification in an unusual format of some very complex issues" Sir Crispin Tickell GCMG, KCVO

"This is an excellent, witty and imaginative book that looks at climate change through the lens of simplicity without the jargon and scientific detail which so often dogs this subject. It has a humour and tongue-in-cheek approach which highlights both the importance of the subject, but the need to engage the reader and not fill him or her with a hopeless sense of guilt. Enjoy." *Tim Smit, CEO Eden Project*

ABOUT THE AUTHOR

Simon Rosser works as a personal injury lawyer in Cardiff. Interests include geography, astronomy, travel, and writing. This is the author's first book, having been inspired to write the book after watching Al Gore's documentary, *An Inconvenient Truth,* in November 2006. All information in the book is based on the latest scientific understanding of the subject sourced from scientific papers and information. As a lawyer, an essential legal skill is the ability to sift through lots of paperwork and information in order to extract the most important, salient points, and this skill has proved invaluable in putting this A-Z guide together from the vast amounts of information and data out there on the subject of global warming. All line illustrations in this book are taken from the author's own original artwork.

Ten per cent of the author's net sale proceeds will be donated to the following global-warming-related organisations:

1 WWF

2 The Alliance for Climate Protection

3 Global Cool

4 Pure, the clean planet trust

FOREWORD

Climate change presents a great challenge for ecosystems and wildlife already ravaged by unsustainable overexploitation, pollution, invasive species, and deforestation. Warming could cause tremendous shifts in globally important regions like the Amazon rainforest and the Arctic. People too, are at risk, but the brutal irony is the places that will bear the brunt of changing climate – low-lying islands and Africa for instance – have contributed least to the problem. Nevertheless there is hope. While humanity is driving rising carbon dioxide levels, we can also devise solutions for reducing them. It's not going to be easy, but we are the best hope for life as we prefer it on this planet. It's all up to us.

Rhett Butler, Mongabay

Rhett Butler has been involved with tropical rainforests since 1995. More importantly, the information sources (peer-reviewed journals, respected researchers, etc.) used by mongabay.com are credible. Furthermore, the site has been praised by a number of well-respected conservation biologists – including Dr Russell Mittermeier of Conservation International, Dr Peter Raven of the Missouri Botanical Gardens, William F Laurance of the Smithsonian Tropical Research Institute, Mark Plotkin of the Amazon Conservation Team, and Dr David L Pearson of Arizona State University, among many others. See <www.mongabay.com>.

INTRODUCTION

It's not possible now to get through the day without hearing something about global warming or climate change on the news, or reading about the subject in the papers.

Indeed to most people, including myself before writing this book, the issues seemed so complex, confusing and contradictory, that it was hard to know exactly what was really going on: whether the world really was warming up, whether the situation was being exaggerated or whether the causes of global warming, if occurring, were manmade or not.

It is now clear that something is indeed going on with Earth's climate. Polar ice caps appear to be melting, or at least the pictures on the news seem to show this to be the case. There is constant mention of rising carbon dioxide levels, and there always appear to be high-powered meetings going on between nations to discuss the climate. One only recently took place in Bali in December where climate change and global warming were hotly (excuse the pun) discussed.

Following a viewing in the cinema in November 2006 of Al Gore's documentary *An Inconvenient Truth*, I felt inspired to research, read about and find out as much as I could on the subject of climate change and global warming. I bought numerous books, scanned the internet and trawled through enormous amounts of information written on the subject, from multiple sources, in order to gain as comprehensive an understanding as I could on the issues of global warming. Information for this book came from organisations such as the National Aeronautics and Space Administration (NASA), The National Oceanic Atmospheric Administration (NOAA), the Stern Review on The Economics of Climate Change, The World Wildlife Fund (WWF), and Mongabay, to name a few. After a while I conceived the idea of an A–Z on the subject, which would make it quick and easy for anyone who wanted a good knowledge of global warming with a book that got straight to the heart of this very complex topic.

All the related issues of global warming and climate change have been packed into A–Z format, with each chapter dealing with a specific point on the subject.

I have used the term 'global warming' as in my view it most accurately describes what is in fact happening to planet Earth on a global scale. However, as a result of global warming, there will of course be climatic change in different parts of the world, whether it be higher temperatures, rainfall, drought or hurricanes, or even perhaps localised cooling.

Global warming: a brief introduction

The term 'global warming' has been in common usage for some time and usually refers to recent warming of Earth's atmosphere, which also implies a manmade or human influence. Each chapter of this book deals with an aspect of it, but as an introduction, here is a quick overview.

Earth's atmosphere comprises many gases: oxygen, nitrogen, carbon dioxide (hereafter abbreviated to CO_2) and water vapour, to name a few. These gases are collectively called greenhouse gases and they keep the Earth's temperature at a comfortable 15°C. Without them Earth would be a chilly -18°C.

Since pre-industrial times, usually taken to be before 1750, we know from ice-core records that CO_2 levels were about 280 ppm, that's 280 parts of CO_2 per million parts of air. As industrialisation got underway humankind started to farm the land more intensely than ever before, which included deforestation for agriculture and settlements. Later – since about 1850 or so – the burning of fossil fuels for energy and transport has added considerably to greenhouse gas levels, particularly CO_2.

This has resulted in CO_2 levels increasing to about 385 ppm, a rise of about thirty-seven per cent from pre-industrial levels – mainly as a result of burning fossil fuels.

How do we know this? Well, data from ice-core records that

go back at least 650,000 years now show us that CO_2 levels have fluctuated naturally during this time between 280 and 300 ppm. CO_2 levels have also been measured accurately from the top of Mauna Loa Volcano in Hawaii since 1958, and results show an increase in CO_2 levels from 315 ppm to 385 ppm since that time. Therefore CO_2 is now at eighty-five ppm more than it has been for at least 650,000 years of Earth's history.

It is a known scientific fact that higher levels of greenhouse gases will lead to higher temperatures, which appears to be happening now. The world has warmed by an average of 0.74 degrees during the last 100 years or so.

As a result of this warming, polar ice has started to decrease and melt, and so have Earth's land-based glaciers. This in turn is causing sea levels to rise, which is putting low-lying islands at risk of flooding or total submersion, and will eventually threaten more and more of the world's coastal cities and regions.

Things may get worse, however, because once Earth's atmosphere starts to warm, the warming itself may cause further positive feedback mechanisms to kick in. A warmer atmosphere holds more water vapour, which is itself a powerful greenhouse gas. This will in turn cause further warming, and so on.

Melting ice results in more sunlight absorbed by the surrounding 'darker' water and land, and that results in further warming, and more melting ice. Methane deposits currently held in a frozen but stable state under the sea and under the permafrost may be released as the oceans warm and permafrost melts, which will cause further warming as methane is a potent greenhouse gas, etc., etc.

That is global warming in a nutshell, but is humankind really to blame? Read this book, and make up your own mind.

I hope to explain everything in an uncomplicated way, and will look at all the issues in much more detail, starting with the involvement of the Amazon rainforest, biofuels, CO_2 and so on, with each chapter having relevance to the phenomenon of global warming. Earth's historical climate is looked at, the sun's role, the Kyoto Agreement is explained, and the findings of the Intergovernmental Panel on Climate Change (IPCC), a unique panel comprising many thousands of researchers and scientists from all over the world, are revealed.

Towards the end of the book, the consequences of global warming are looked at in terms of the weather, diseases, extinction, and importantly what the world and every human being on it can do to try and alleviate the crisis that planet Earth and all of us are facing.

I hope from reading this book you will learn something and perhaps get concerned enough to do something about the problem...

A

AMAZON

We start our A–Z journey on global warming with the Amazon rainforest, which has an incredibly important role to play in maintaining balance in the Earth's climate, in ways that are only just being understood. The Amazon is inextricably linked to the issue of global warming and therefore a very good place to start our inquiry into what may be the biggest threat to our existence on this planet.

Amazon facts

The Amazon river basin contains the largest rainforest on Earth and covers approximately forty per cent of the South American continent. The rainforest is located in eight countries. Brazil has sixty per cent, with Colombia, Peru, Venezuela, Ecuador, Bolivia, Guyana, Suriname and French Guyana between them containing the rest.

The Amazon forest is a natural reservoir of genetic diversity, containing the largest and most species-rich tract of tropical rainforest that exists. The Amazon contains an amazing thirty per cent of Earth's species. One square kilometre can sustain about 90,000 tons of living plants! It's also amazing to consider that one in five of all the birds in the world make the rainforest their home.

The Amazon basin is drained by the Amazon River, the world's second longest after the Nile. The river is essentially the lifeline of the forest. It is the most voluminous on Earth and its *daily* freshwater discharge into the Atlantic is enough to supply New

York City's freshwater needs for nine years![1] New measurements recently taken by scientists, however, suggest that the Amazon may actually be the longest river in the world. No doubt this will be confirmed if true, at some point in the future!

A few thousand years ago tropical rainforests covered as much as twelve per cent of the Earth's land surface, but today the figure is below five per cent. The largest stretch of rainforest can be found in the Amazon river basin, over half of which is situated in Brazil.[2]

Why is the Amazon so important in the context of global warming?

The rainforest acts as a major store of carbon and produces enormous amounts of oxygen. The Amazon has been referred to as 'the lungs of the Earth' because of its affect on the climate. The way this is achieved is of course through photosynthesis, the process by which green plants and trees use the energy from sunlight to produce food by taking CO_2 from the air and water and converting it to carbon. The by-product of this is oxygen.

The Amazon therefore helps recycle CO_2 by turning it into oxygen, and it is estimated that the Amazon produces about twenty per cent of this essential gas for Earth's atmosphere.

Trees, plants and CO_2

Levels of CO_2 in the atmosphere have been measured since 1958, from a monitoring station located on Mauna Loa volcano in Hawaii. They show sharp annual increases and decreases in CO_2 levels, similar to the tooth on a saw. The readings seem to mimic a breath of air being taken in and out, almost as if the Earth is breathing. They correspond to the amount of vegetation on the planet (most of which is in the northern hemisphere, as the landmass there is greater), taking in CO_2, and giving out oxygen. During summer in the northern hemisphere, when the Earth is tilted towards the sun, Earth's vegetation is able to photosynthesise, resulting in an uptake of CO_2, causing world-wide CO_2 levels to drop. In winter in the northern hemisphere, when Earth's axis is tilted away from the sun, the opposite happens, causing CO_2 levels to rise again.

When one becomes aware of the correlation between the Earth's vegetation and CO_2 levels, it is easy to understand why the Amazon, and rainforests in general, are such an important part of Earth's ecosystem.

The problem is, however, that although the measurements taken at the volcano in Hawaii show sharp up and down annual readings, the measurements also show a simultaneous steady upward trend in CO_2 levels. The importance of CO_2 in relation to global warming will be a recurring theme throughout this book, and will be looked at further in Chapter C.

What has been happening in the Amazon?

A worrying trend is the Amazon having experienced two consecutive years of drought, in 2005 and 2006.

The drought in 2006, which left rivers dry, stranded thousands of villagers, and put regional commerce at a standstill, was the worst on record.

A second year of drought is of great concern to researchers studying the Amazon ecosystem. Field studies by the Massachusetts-based Woods Hole Research Centre in the USA, suggest that Amazon forest ecosystems may not withstand more than two consecutive years of drought without starting to break down. Severe drought weakens forest trees and dries leaf litter leaving forests susceptible to land-clearing fires set during the July-October period each year. According to the Woods Hole Research Centre, it also puts forest ecosystems at risk of shifting into a savannah-like state.[3]

A recent experiment carried out by a team of researchers suspended 5,600 large plastic panels between 1 and 4 metres (3.2–13.1 feet) above the ground to mimic severe drought conditions, where as much as eighty per cent of a one-hectare plot is deprived of eighty per cent of rainfall. Measuring rainfall, soil moisture, leaf and canopy characteristics over time, it was found that after four years the rainforest trees began to die while leaf litter dried and became tinder for wild fires.[4]

Another factor is the El Niño Southern Oscillation (ENSO)

event, a climatic phenomenon that influences much of the climate in the region, particularly Northeast Brazil, and the northern Amazon. ENSO brings with it dry conditions in the above areas, and manmade climate change is thought to increase this naturally occurring phenomenon in the future. ENSO is further looked at in Chapter W.

Some climate models have suggested that temperatures in the Amazon may increase by 2 to 3°C (3.6–5.4°F) by the year 2050, together with a decrease in rainfall during the dry period.

If the drought continues, based on the results of the aforementioned experiment, 2007/8 could be a turning point for the forest, which may mean that a tipping point will be reached where the forest will start to die, with catastrophic consequences for Earth's climate.

If this trend continues, according to the WWF, between thirty and sixty per cent of the Amazon rainforest could become dry savannah, rendering the forest a source of CO_2 instead of a sink/ store of it, which it currently is.

There are ways in which we can all help try and sustain this vast and ecologically important expanse of rainforest, and these will be discussed in Chapter Y.

The Amazon will be further considered in Chapter D, where the problem of deforestation is looked at.

We will now consider the importance of biofuels as an alternative source of fuel, and how biofuels may help in the fight against global warming. Ironically, this is also causing problems for the Amazon and other rainforests, as areas of forest are cleared for the planting of crops for biofuel production.

Key points

- ➢ The Amazon rainforest contains about thirty per cent of Earth's species.
- ➢ World rainforest cover has over thousands of years decreased from twelve per cent to five per cent.

- The Amazon helps to recycle CO_2, a gas which contributes to global warming and while doing so produces about twenty per cent of Earth's oxygen.
- CO_2 levels rise and fall with the seasons. There is greater landmass and hence vegetation in the northern hemisphere, which means that when Earth is tilted towards the sun during northern summertime, CO_2 levels drop as a result of there being greater uptake of CO_2 from photosynthesis. During the winter, the opposite happens and CO_2 levels rise again.

---

1 Mongabay, <www.mongabay.com>.
2 Ibid.
3 Ibid.
4 Ibid.

Above: Trees and plants photosynthesise. During the light dependent-stage, complex chemical reactions occur whereby light energy, CO_2 and water produce sugar (food) and oxygen.

Below: During the light-independent stage trees and plants capture CO_2 from the atmosphere.

B

BIOFUELS

'Biofuels' is undoubtedly a word that will be heard a lot more often over the coming years, but what are biofuels? Where do they come from? And what is their significance in relation to global warming?

Biofuels can be described as encompassing any fuel that is derived from biomass, i.e. living organisms or their metabolic by-products. For example,

Crops such as corn
Dung from living animals

Although there is still something of a scientific debate going on over the advantages of biofuels, it is thought that the main advance over fossil fuels (coal, oil and gas) is that burning them to release energy does not cause a net increase of CO_2 levels in the atmosphere. This is because the sources of biofuels, crops for example, have already taken a corresponding amount of CO_2 out of the atmosphere during their growth cycle when they photosynthesise. When this occurs, plants/crops release oxygen and retain the carbon to use as energy.

The carbon is then released when the crop is eventually burnt in order to release its energy. As long as new crops are planted in place of the ones that are burnt, there will be no overall increase in the amount of CO_2 released into the atmosphere. So, while crop-based biofuels don't reduce the amount of CO_2 in the atmosphere, they are thought to be more or less carbon neutral.

The difference with fossil-fuel deposits, such as coal, is that the coal deposits have been formed in the earth over millions of years and are therefore considered to be energy deposits rather than part of the energy cycle. The burning of fossil fuels on a scale required to satisfy humankind's energy needs, over a relatively short period of time, hundreds of years as opposed to the millions of years it has taken the deposits to form, means that the burning of such fuels adds considerably to the levels of CO_2 in the atmosphere. This in turn adds to the greenhouse gases already present in our atmosphere, and contributes to the warming of the Earth's climate.

Forms of biofuel

Biofuels can be in either solid or liquid form.

Solid biofuels such as wood or even manure (dried cow dung) can be burnt to heat water, which can then be used to power a turbine, which can generate electricity. Liquid biofuel in the form of ethanol can be derived from crops.

Biofuel uses

The major benefit probably comes from liquid biofuel, for the creation of ethanol or biodiesel. Ethanol is a substitute for fossil-fuel-based petrol, and biodiesel is diesel made with crops in place of oil, which is a replacement for traditional diesel fuel in diesel motor vehicles. While diesel cars are more fuel efficient than their petrol counterparts, biodiesel vehicles produce even less CO_2. Neither is as efficient, however, as vehicles running on mostly ethanol-based fuels.

To run on fuel that has a greater than ten-per-cent mixture of ethanol, however, vehicles need a flexi-fuel modified engine.

The USA grows mainly corn crops, which can be converted to ethanol. In Brazil sugar cane is grown, and in the UK rapeseed is used.

In fact Brazil is the world's leading producer of biofuels, and the flexi-automobile engine that can use it.[1]

Recent legislation in the UK and the USA will ensure that petrol contains a certain percentage of ethanol. The introduction of the Renewable Transport Fuels Obligation in the UK (RFTO), expected by 2008, could reduce CO_2 emissions by about 1,000,000 tonnes per annum, by introducing ethanol into the transport sector, and also by mixing ethanol with traditional fuel that can be used in unmodified vehicles.[2]

On a recent visit to my local Cardiff (South Wales) Morrisons superstore to buy fuel, I noticed they already had in place a lone E85 ethanol fuel pump. It probably isn't getting too much use at the moment, but it won't be long before all the major gas companies follow suit.

Back in Brazil, however, things appear to be well ahead of the game, as over fifty per cent of cars sold in 2005 were flex-fuel cars, which can run equally well on either ethanol or petrol!

Brazil was spurred on by the country's desire to be less dependent on foreign oil during the 1973 oil crisis, something that the USA has also now decided to do by ensuring its fuel has a twenty per cent biofuel mixture by 2017.

The USA recently passed a law – in August 2005 – called the Energy Policy Act, which sets objectives to diversify the USA's energy supply, in order to become less dependent on foreign oil, and to develop cleaner energy technologies. Part of the plan is to develop new biorefineries for the production of biomass, bioproducts and biomass-based heat and power.[3]

Environmentally friendly or not?

Despite the benefits of using biofuels, there is a drawback. This is in the amount of land required to grow the crops necessary to produce the biofuel in the first place. There are already concerns that vast tracts of tropical rainforest such as the Amazon in Brazil are being cleared to plant sugar cane and other crops for this kind of production. Another problem is the cost of corn, which happens to be an essential ingredient of basic foodstuffs, which is escalating, causing further problems for the poorest people on the planet. Clearly, it would be counter-productive if a situation were to develop where the CO_2-absorbing tropical rainforests were destroyed to plant crops to turn into CO_2-friendly biofuels!

There is also concern that as a by-product of growing corn or other crops used for biofuel production, environmental damage is caused by the fossil-fuelled tractors, processes, fertilisers and so on used in the growing process, which is not a truly carbon-neutral position at all. Recently scientists have revealed that some crops would have to be grown for centuries just to repay the amount of carbon released into the atmosphere by their initial cultivation. It would seem that only certain crops grown in certain circumstances remain truly carbon neutral. If rainforest has to be cleared first to plant them, then the loss of the crucial carbon-storing rainforest would mean that any crops grown for biofuels would not be carbon neutral at all – in fact, quite the opposite!

Green grasses

A possible answer to this may have been found by a team from the University of Minnesota in the USA, led by a Mr David Tilman. The team has discovered that prairie grasses are far more effective at yielding energy than traditional cornfields. More importantly, it found that the grasses actually removed far more CO_2 from the atmosphere while growing, even taking into account the harvesting and transportation process necessary to turn them into biofuels, rendering them truly carbon neutral when burnt for fuel.

Biofuels such as ethanol will become part of everyday use as a blend with traditional fuel in the transport sector, which will help reduce CO_2 greenhouse gas emissions, which in turn will go some way to reduce the impact the transport sector has on global warming in the future.

Vehicle manufacturers such as Ford and Saab have both recently introduced vehicles for the UK market that can run on a mixture of E85 fuel, significantly reducing the CO_2 emissions produced.

When coming across fuel at petrol stations, you will find the description 'E10' or 'E85'. The former denotes a ten per cent biofuel mix, the later eighty-five. As mentioned, you would need a flex-fuel car to use E85, otherwise you may damage your engine!

Even the large oil companies are taking the biofuels boom seriously. British Petroleum (BP) has recently invested millions in a new biofuel research programme in India. Further information can be found at BP's website.[4]

According to the WWF, worldwide biomass could deliver nine per cent of global primary energy and twenty-four per cent of electricity requirements by the year 2020. Biomass use in combined heat and power production systems is the most efficient.

As we will see in Chapter E, however, the development of alternative technologies will at some point eclipse the traditional combustion engine used to power modern motor vehicles, together with the fossil/biofuels used to run them.

We now move on to Chapter C and look at the role of the gas CO_2, which is perhaps the greatest manmade contributor to global warming, caused primarily by humankind's insatiable appetite for fossil fuels for energy production.

Key points

- Biofuels are fuels from biomass, living organisms or their metabolic by-products.
- Ethanol is liquid biofuel and can be used in place of traditional fossil-based fuels.
- Biofuel from prairie grasses appears to be truly carbon neutral, meaning their use does not result in a net increase of CO_2 in the atmosphere.

---

1 Stern Review on The Economics of Climate Change, Part VI.

2 Department of Transport, <www.dft.gov.uk>.

3 US Department of Energy.

4 British Petroleum, <www.bp.com>.

Above: Crops such as corn are grown for use in biofuel production. Sugarcane and rapeseed are also used.

Below: The corn contains starch that can be converted to sugar. This is fermented to create ethanol and CO_2. The ethanol is then purified so it can be used as fuel.

C

CARBON DIOXIDE

Okay, so we are now well into our alphabetic A–Z journey through global warming. C for Carbon Dioxide is one of the main players in the global-warming problem. Carbon dioxide, chemical symbol CO_2, is a chemical compound composed of one carbon and two oxygen atoms.[1]

CO_2 is present in the Earth's atmosphere at a low concentration, about 0.038 per cent by volume, and is one of many gases that make up Earth's atmosphere (see Chapter G). CO_2 is measured in parts per million by volume of air (ppmv). Atmospheric CO_2 derives from many natural sources, including volcanic eruptions, the combustion of organic matter, the respiration of living aerobic organisms, and unfortunately from manmade (anthropogenic) sources, which we all know from the news is being linked to global warming and climate change.

Since the Industrial Revolution, particularly the mid-nineteenth century, the burning of fossil fuels for energy to provide electricity, power factories and homes, and for all our transport needs, has released massive amounts of CO_2 into the atmosphere. Not only the burning of fossil fuels, but changes in the use of the land for agriculture and deforestation (looked at in the next chapter), have further added to global manmade CO_2 levels.

According to the WWF some twenty-nine gigatons, which is 29,000,000,000 metric tons of CO_2, were, in 2004 alone, added to the atmosphere from burning coal, oil and gas.

If we go back 250 years or so, to pre-industrial times, usually taken to be approximately 1750, CO_2 levels in the atmosphere stood at about 280 parts per million by volume (ppmv). However, levels of the gas have been increasing steadily ever since.

How do we know this?

Well, pioneering scientist Charles Keeling (1928–2005) started taking atmospheric CO_2 measurements in 1958 from Mauna Loa volcano in Hawaii. Those measurements have been recorded and are now known as the Keeling curve. Charles Keeling was the professor of oceanography at the Scripps Institute of Oceanography (SIO), in San Diego, USA. He followed the work of another eminent scientist and director of the SIO, Roger Revelle. Dr Revelle was instrumental in creating the Geophysical Year in 1958, and SIO's first programme looking at atmospheric CO_2 back in 1956.

Monthly CO_2 measurements were collected from a height of 3,397 metres (11,140 feet) at the Mauna Loa Observatory situated on the slopes of Earth's largest volcano, Mauna Loa, which was chosen for its remoteness from populations and vegetation, so as not to skew the readings.

Measurements have been taken over a fifty-nine-year period, between 1958 and present, and show an increase in CO_2 levels of 70 ppmv from about 315 ppmv to approximately their current level of 385 ppmv. The effects of CO_2 in the atmosphere can even be measured on a cyclical basis, and this can be seen in the saw-toothed Keeling graph shown over the page. Because there is a greater land area, and thus far more plant life in the northern hemisphere (as mentioned in Chapter A) compared to the southern hemisphere, there is an annual fluctuation of about five ppmv peaking in May and reaching a minimum in October. This corresponds to the northern hemisphere growing season. The amount of CO_2 in the atmosphere drops towards spring, when uptake by the plants and trees by photosynthesis is greatest. The opposite occurs in winter when the plants die off and CO_2 levels rise again.[2]

Continuous readings in this way have been taken only since 1958. However, scientists have discovered that prior to the industrial era, circa 1750, CO_2 levels stood at about 280 ppmv, and this data has been revealed from air trapped in ice core records, taken from both the Antarctic and Arctic.[3]

Perhaps most startling is the fact that CO_2 levels are now about eighty-five ppmv higher than at any time during the last 650,000 years. Records from ice-core records go back that far and have shown atmospheric CO_2 levels to range from 180-300 ppmv during that period. The level of CO_2 in our atmosphere now stands at 385 ppmv, and is increasing steadily.[4, 5, 6]

The Keeling curve has become one of the most recognisable images in modern science, as it shows with no uncertainty the effects of humankind's fossil-fuel pollution of Earth's atmosphere.

CO_2 levels have increased by thirty-seven per cent since pre-industrial times and have been increasing by an average of almost 1.4 ppmv a year since measurements began in 1958 – although some months the figure has been higher, sometimes lower. In the last ten years, the average increase appears to be about 1.9 ppmv each year, which indicates the rate of increase is increasing. This is looked at further in Chapter I.

Where does all the CO_2 go?

It is estimated that about fifty per cent is absorbed by the oceans and land (soil, plants, trees, etc.) in equal amounts, and fifty per cent remains in the atmosphere. The oceans absorb vast amounts of CO_2 and act as a major sink/store for the gas, just as do the forests of the Amazon. However, the oceans take a relatively long time to absorb the CO_2 that is pumped into the atmosphere, and therefore the effects of current CO_2 levels may not be reflected by the oceans for some time to come. The oceans can sustain many times more CO_2 than the atmosphere can. According to NOAA the oceans have taken up about 118,000,000,000 metric tons of CO_2 from human sources (anthropogenic CO_2) between 1800 and 1994. This equates to about forty-eight per cent of all manmade CO_2, which would be enough to push atmospheric CO_2 up by an additional fifty-five ppm.

Why is carbon dioxide such a problem?

Basically global-warming theory predicts that increasing amounts of CO_2 (and other gases) in the atmosphere tend to enhance the greenhouse effect and thus contribute to global warming. Despite CO_2 being present in the atmosphere in small concentration, natural CO_2 levels are a very important component of Earth's atmosphere. As mentioned earlier, CO_2 is one of Earth's natural greenhouse gases and it helps the Earth maintain its temperature by trapping some of the sun's heat, which would otherwise escape back into space. If this did not happen the Earth would be some 30°C (54°F) cooler and have an average temperature of about -18°C (-0.4°F) – pretty chilly, unless of course you are a penguin!

CO_2 is also essential for life on Earth. Photosynthesis, the process by which plants and trees absorb CO_2 and produce oxygen, could not occur without it. In the distant past volcanoes were the main source of Earth's CO_2, and there are still lots of active volcanoes on Earth, such as Mount Etna and Stromboli in Italy, which have been erupting continuously for thousands of years. Erupting volcanoes are just part of Earth's natural CO_2 cycle, and the CO_2 they emit will eventually be absorbed back into the oceans and the land.

CO_2 is only one of the gases that make up the Earth's atmosphere that are collectively referred to as greenhouse gases. As we shall see in later chapters, the higher the level of greenhouse gases of which manmade CO_2 is a component, the higher the Earth's temperature is likely to be.

The effects of higher temperatures could be catastrophic, as we shall be reminded throughout this book.

We will now look at deforestation, which is a continuing problem, and which destroys the Earth's rainforests' ability to soak up CO_2. The rainforests' destruction also adds to CO_2 levels as dead and decaying trees release their stores of carbon back into the atmosphere that were taken out over many decades of growth.

Key points

- CO_2 is just one of Earth's greenhouse gases and makes up just 0.038 per cent by volume of atmospheric gas.
- Levels of CO_2 have increased from 280 to 385 ppmv since circa 1750, an increase of thirty-seven per cent, mainly as a result of burning fossil fuels.
- CO_2 is a global-warming gas and current levels are higher than at any time in the last 650,000 years.
- Professor Charles Keeling started taking measurements of CO_2 from Mauna Loa Observatory in Hawaii, and they show an increase from 315 to 385 ppm since 1958.

---

1 Wikipedia, <www.wikipedia.org> (carbon dioxide).
2 NASA, <www.visibleearth.nasa.gov>, the Keeling curve.
3 Stern Review on The Economics of Climate Change, Part I.
4 Ibid.
5 Real Climate, <www.realclimate.org>.
6 Mongabay, <www.mongabay.com>.

Above: Prior to the Industrial Revolution, circa 1750, atmospheric CO_2 levels stood at about 280 parts per million (ppm) in air.

Below: 250 years of industrialisation, particularly from the mid-nineteenth century, have resulted in global CO_2 levels increasing to 385 ppm, mainly as a result of burning fossil fuels.

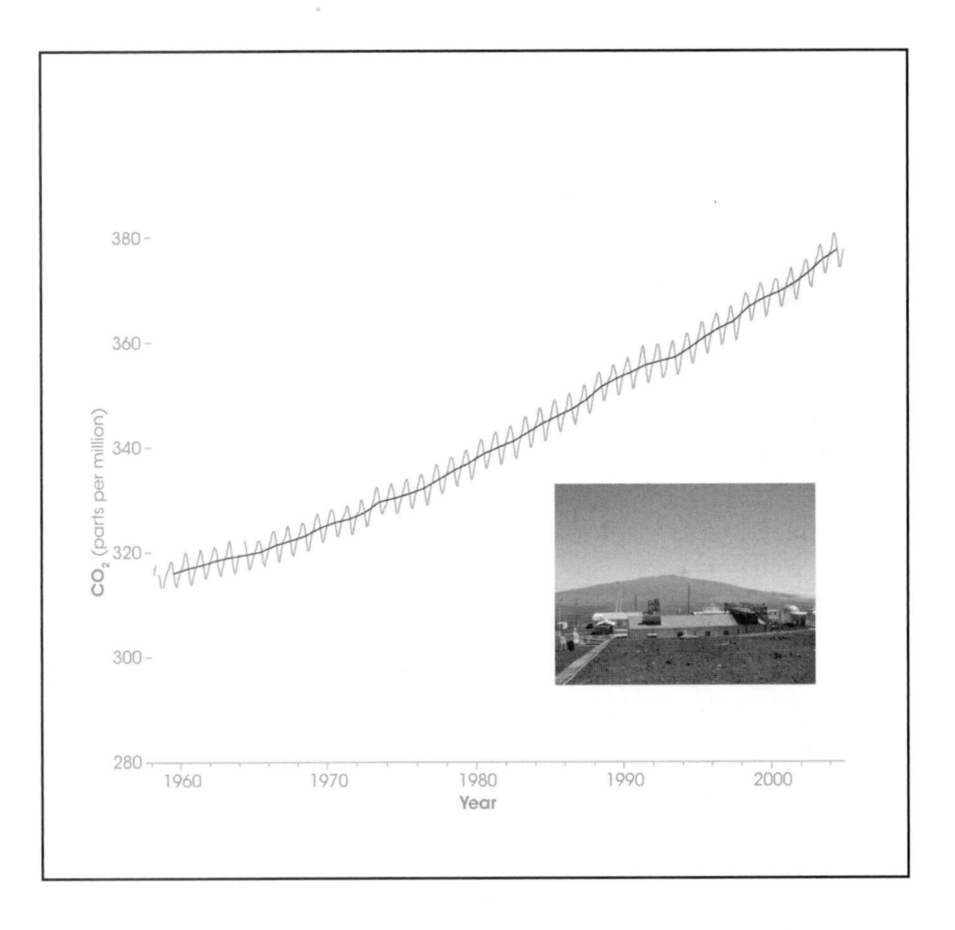

The Keeling Curve showing CO_2 concentration taken from Mauna Loa in Hawaii.

Credit NASA Visible Earth, <http://visibleearth.nasa.gov>.

D

DEFORESTATION

Deforestation is basically the loss or destruction of forest habitat, primarily as a result of the action of human beings.

It is the single largest source of land-use greenhouse gas emissions. It accounted for about eighteen per cent of emissions in the year 2000 globally[1] (most recent year that land-use-change greenhouse-gas emissions figures are available).

As we know from Chapter A, trees and vegetation act as sinks or stores for CO_2, one of the most important greenhouse gases. Stored carbon taken out of the atmosphere by photosynthesis through decades of growth is released back into the atmosphere as vegetation and trees are cut down and burnt, or, as unburned organic matter slowly dies. This process adds to atmospheric CO_2 levels.

Forest classification

The Food and Agriculture Organisation (FAO) of the United Nations (UN), which is the leading source of information on the status of the world's forests, define forests as 'land with a tree canopy of greater than 10%, and an area of more than half a hectare'. The organisation defines deforestation as 'the conversion of forest to another land use or long term reduction of the tree canopy cover below the minimum 10% threshold'.[2]

Land change and CO_2

Changes in the use of land account for eighteen per cent of global greenhouse gas emissions, with emissions from agriculture producing a further fourteen per cent, and waste three per cent, meaning total non-energy emissions – i.e. not from burning fossil fuels – accounts for thirty-five per cent, based on year 2000 figures. If we are just talking about CO_2, rather than general greenhouse gases, then land-use change was responsible for about 22.7 per cent of emissions in year 2000.

Land-use changes are driven almost entirely by emissions caused through deforestation, which is highly concentrated in a few countries. Indonesia contributes approximately thirty per cent of land use to CO_2 emissions, with Brazil at about twenty.

While land-use emissions are predicted to fall by 2050, this is due to the fact that it is assumed that a country will stop deforesting only after eighty-five per cent of its forests have been cleared![3]

This is clearly a frightening prospect.

Disappearing forests

It is estimated that about 80,000 acres or 32,000 hectares are being lost every day.[4] This is the equivalent of about 117,000 square kilometres (45,173 square miles) each year.

Total world rainforest cover is about 6,000,000 square kilometres (2,316,612 square miles), and this equates to about five per cent of Earth's land surface.[5] Only a few thousand years ago, rainforests covered about twelve per cent of the world's land surface, about 15,500,000 square kilometres, (6,000,000 square miles). A quick calculation reveals that if forest cover is lost at the rate of 117,000 square kilometres a year, then it would take only approximately fifty-one years for it to be destroyed! (6,000,000 divided by 117,000.)

Indeed, the Stern Review on The Economics of Climate Change reports that 'If the current rate of deforestation continues, the top ten deforesting nations would completely clear their forests before 2100'.[6]

Destruction at this level would lead to the release of vast

amounts of CO_2 into the atmosphere, further thickening the CO_2 'blanket' that surrounds our planet, and no doubt lead to an increased warming of the atmosphere.

Between 2000 and 2006 Brazil lost nearly 150,000 square kilometres (57,915 square miles) of forest, an area the size of Greece, and since 1970 over 600,000 square kilometres (231,660 square miles) have been destroyed.

It is now estimated that almost twenty per cent of the Amazon has been destroyed, which is alarming when one considers that the Amazon rainforest represents almost half of the world's tropical rainforests.[7]

Rainforest destruction

As mentioned, the main reason is land-use change. Deforestation in Brazil is closely related to the economic health of the country. A slowdown in economic growth between 1988–1991 matched a decline in deforestation, while a rapid increase in growth during 1993–1998 saw the opposite.[8]

The main causes of deforestation are as follows:

1 Cattle ranching

Almost sixty per cent of deforestation is caused by large-scale cattle ranching, which in turn is partly being driven by Europe's appetite for processed meats. Between 1990 and 2001 the percentage of Europe's meat imports that came from Brazil increased from forty to seventy-four per cent, according to the Centre for International Forestry Research.[9]

2 Activities of farmers

Thirty per cent occurs from the activities of farmers encouraged by the government to settle on forestlands through policies that give them rights to continue to use the land after they have claimed it as their own. Thereafter farmers usually employ the slash-and-burn method to clear the forest for settlement/use.

3 Fires, mining and road construction

Virtually all forest is cleared using fire, and about three per cent of the Amazon is cleared for the purposes of mining and new construction. In 1987 some 500,000,000 tonnes of CO_2 were released, together with other polluting particles, into the atmosphere.

4 Logging, legal and illegal

Loggers tend to remove the oldest and most valuable trees, releasing their stored carbon, as well as that of some neighbouring trees damaged in the process.[10] Logging accounts for between two to four per cent of forest destruction.

If logging were undertaken in a controlled fashion, with re-planting and time left for forest re-growth, then CO_2 emissions could be offset over time.

5 Large scale commercial agriculture

Businesses are moving in using the land to produce crops for biofuels, as discussed in Chapter B, which destroys another one to two per cent of the forest.

The forest is also being cleared in order to plant soya beans, the growth of which is being driven by Western demand for an increase in supermarket availability of pork, chicken and beef. Soya is increasingly used as a food source for livestock, which is mixed with traditional maize as feed for farm animals.

Sharing blame

It's not entirely fair to blame developing nations for all the deforestation, however. While countries like Brazil and Indonesia may be the main culprits now, up until the early twentieth century emissions of CO_2 through land-use changes came from developed nations. It's a natural step for developing nations to clear forestland for agriculture and habitation. The fact is that as developed nations have already deforested many areas long ago, there is more pressure on developing nations to preserve what is left. Of course, population growth is another major factor, which will be discussed in a later chapter.

Another significant point is that trees in tropical forests typically hold on average about fifty per cent more carbon per hectare than trees outside the tropics.[11] Therefore deforestation in these areas causes greater amounts of CO_2 to be released into the atmosphere than deforestation outside of the tropics.

Future of the forests

Remarkably, when talking about land-use-change emissions, countries and political groupings such as the USA, Europe and China were, in the year 2000, net absorbers of CO_2, owing to their aforestation (planting new forests) and reforestation (re-establishing old forest areas) programmes. However, the planting of one tree does not offset the damage caused by the removal of another, as trees absorb CO_2 very slowly. It could take 100 years for a growing tree to recover all the CO_2 released when a mature tree is cut down![12] For this reason, carbon offset programmes that recommend planting tress are pretty worthless, due to the time it takes to remove CO_2 from the atmosphere.

There is some good news, however. In 2006 the Brazilian government announced a sharp drop in deforestation. Loss for the year 2005/6 was 13,100 square kilometres (5,057 square miles), down more than forty per cent from the year before.[13]

It's too early to say whether this is a declining trend, or just one good year out of the previous eight when deforestation levels were all in excess of 16,000 square kilometres (6,177 square miles). A very recent Mongabay report suggests the latter as it seems deforestation in the Brazilian Amazon rose sharply in the second half of 2007 as a result of surging prices for beef and grain, according to a top Brazilian environmental official.

As the world's forests are being destroyed, huge amounts of CO_2 are being released back into the atmosphere. The forests that once acted as stores as a result of absorbing CO_2 will no longer be standing. That will push CO_2 levels higher, thereby contributing to the warming of Earth's climate.

We will now look at ways in which we can reduce our reliance on fossil fuels in the next chapter, Electric Vehicles and Transport.

Key points

- Deforestation is the largest source of land-use greenhouse-gas emissions, contributing up to eighteen per cent of global greenhouse gas emissions, 22.7 per cent of CO_2 emissions in the year 2000 (most recent available figures).
- Approximately 117,000 square kilometres of rainforest are destroyed each year.
- Approximately twenty per cent of the Amazon rainforest has already been destroyed.
- Thousands of years ago, rainforests covered about twelve per cent of Earth's land surface. As a result of deforestation rainforest cover has reduced to about five per cent.

---

1 Stern Review on The Economics of Climate Change, Annex 7f.
2 Mongabay, <www.mongabay.com>.
3 Stern Review on The Economics of Climate Change, Part III.
4 Mongabay, <www.mongabay.com>.
5 Ibid.
6 Stern Review on The Economics of Climate Change, Part III.
7 Mongabay, <www.mongabay.com>.
8 Ibid.
9 Mongabay, <www.mongabay.com>.
10 Stern Review on The Economics of Climate Change, Part III.
11 Ibid.
12 Ibid.
13 Mongabay, <www.mongabay.com>.

Above: Rainforests are destroyed for a variety of reasons, including logging, cattle ranching and commercial agriculture.

Below: Where rainforests once stood, motorways and settlements replace them, meaning that the rainforests' ability to capture CO_2 *and produce* O_2 *is lost, which will affect Earth's climate.*

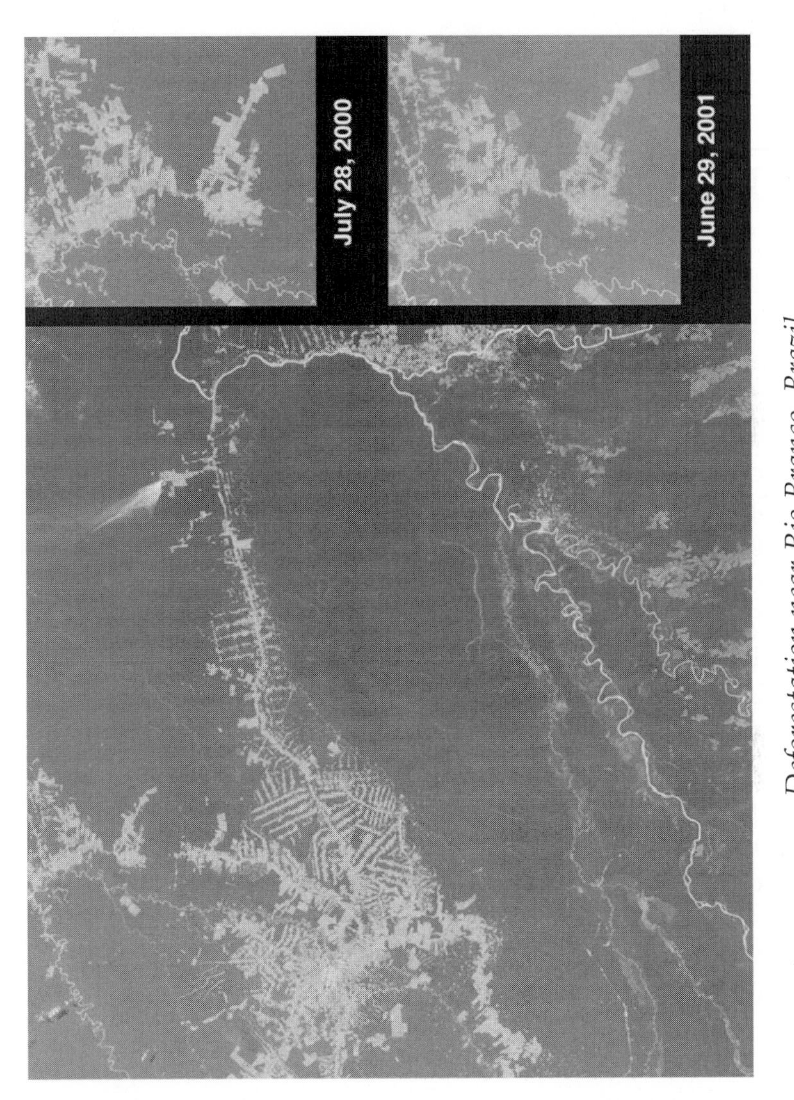

Deforestation near Rio Branco, Brazil.
Settlement and deforestation surrounding the Brazilian town of Rio Branco are shown here.
Credit NASA Visible Earth, <http://visibleearth.nasa.gov>.

E

ELECTRIC VEHICLES AND TRANSPORT

While on a Virgin Atlantic flight recently (my share of CO_2 from the flight was offset! Please see Chapter Y), I decided to watch one of the onboard documentaries, *Who Killed the Electric Car?*

The documentary reveals that during the late 1990s General Motors (GM) produced an electric car, which was called the EV 1. Some 800 or so were available on three-year leases. Owners of these almost silent, zero-CO_2-emission vehicles seemed very happy with their cars, particularly as they were being driven around California, in the USA, which has the strictest CO_2 emission regulations in that country.

When the leases expired on the vehicles, GM recalled all their vehicles and impounded them in a secure area. After they had been sitting there for some considerable time, the previous lease-owners managed, collectively, to raise over $1,000,000 to purchase the cars from GM, who turned down their offer.

The cars were eventually taken away to the desert and crushed! This also happened to some perfectly respectable-looking electric Toyota RAV 4s.

The documentary explores the role of the automobile manufacturer, the US government, the oil industry and the apparent lack of public interest as reasons for the destruction of the vehicles. Members of the public were interviewed in the documentary, who like myself had no idea that electric vehicles were available and driven around the streets of California producing zero global-warming greenhouse gasses!

So, what has happened since then, and what relevance do electric vehicles and other forms of transport play in the global-warming problem?

Transport and emissions

Transport accounts for about fourteen per cent of global greenhouse gas emissions, which can be further broken down by mode of transport, as follows: seventy-six per cent from the road, twelve per cent from aviation, ten per cent from shipping, and two per cent from rail.[1]

According to the Stern Review the largest source of transport emissions comes from North America, about thirty-seven per cent in total.

Emissions from aviation in particular are becoming a major concern as they have an amplification effect on global warming owing to the gasses from burning fuel at high altitudes. Between 2005 and 2050 emissions from the aviation sector are expected to grow faster than other forms of transport, with a tripling over the period compared to a doubling from road transport emissions.[2]

The WWF states that,

> 'Globally, aviation accounts for approximately 4 to 9 per cent of the climate change impact of human activity. In the EU it is higher, currently contributing between 5 and 12 per cent. The EU accounts for approximately half of the CO_2 emissions from international aviation reported by developed countries.'[3]

Aeroplanes produce mainly CO_2 and water vapour (itself the most potent greenhouse gas) when they fly. 'Con trails', short for 'condensation trails', are the long white trails left behind by the aircraft as it flies through the sky, and these are caused by water vapour emitted from the aircraft's engines.

It certainly looks like atmospheric pollution from aircraft is set to get a whole lot worse in the future as airports all over the UK, and the world, are expanding at alarming rates, to cater for the desire for cheap short- and long-haul travel.

Developing countries such as India and China are now beginning to experience the same sort of rapid increase in cheap air travel that has swept across much of the western world. Why should someone living in a country such as India take a five-day bus ride across country when he or she can hop on a budget airline and cut the travel time by days? As economies grow in these countries, there will be a rising middle-class able to afford air travel, which in turn will push up demand for flight, and with it increased CO_2 emissions. China is estimated to build about thirty-seven new airports in the next five years or so!

Further recent research has suggested that global CO_2 emissions from shipping, thought to be about 1.4 per cent, may be higher than previously thought, perhaps up to 4.5 per cent of global emissions, which is more than double that of aviation. This isn't surprising when one considers that almost ninety per cent of freight is transported by sea.

So, the answer seems to be for further technological advances in the transport sector by developing electric vehicles, vehicles that run on biofuels, carbon-neutral transportation and ultimately biofuel-burning jet engines for planes.

Battery and biofuel power

This appears to be starting to happen, now! After feeling a little confused about the decision of GM to destroy all their electric vehicles, it seems that behind the scenes a refinement of the technology developed for their EV 1s, which were unceremoniously crushed, has been ongoing. GM have recently revealed their new Chevrolet Volt concept car, which appears to be far more advanced and practical than the EV 1s developed in the late 1990s.

GM's Chevrolet Volt is a battery-powered, four-passenger electric vehicle powered by GM's E-flex system, their next-generation electric-propulsion system. The vehicle has a one-litre three-cylinder turbo-charged engine, which spins at a constant speed, creating enough electricity to replenish the battery.

The Volt draws from GM's previous experience when it launched the EV 1 in 1996. According to GM Vice Chairman Robert A Lutz,

> 'The EV 1 was the benchmark in battery technology and was a tremendous achievement, even so, electric vehicles in general had limitations. They had limited range, limited room for passengers or luggage, couldn't climb a hill or run the air conditioning without depleting the battery, and had no device to get you home when the battery's charge ran low.
>
> 'The Chevrolet Volt is a new type of electric vehicle. It addresses the range problem and has room for passengers and their stuff. You can climb a hill, turn on the air conditioning and not worry about it.'

The Volt can be fully charged in six hours by plugging it into a 110-volt outlet. When the lithium-ion battery is fully charged, the Volt can be driven more than sixty city kilometres of pure electric range. When the battery is depleted the engine kicks in to replenish the battery once again.

In addition the vehicle is designed to run on E85, a fuel blend of eighty-five per cent ethanol and fifteen per cent gasoline, thus reducing reliance on fossil-fuel propulsion.

Larry Burns, GM's Vice President for research and development and strategic planning, states:

> 'Today there are more than 800 million cars and trucks in the world. In 15 years' time that will grow to 1.1 billion vehicles. We can't continue to be 98 percent dependent on oil to meet our transportation needs. Something has to give. We think the Chevrolet Volt helps bring about the diversity that is needed. If electricity met only 10 per cent of the world's transportation needs, the impact would be huge.'

According to Ed Welburn, Vice President, Global Design,

> 'First and foremost, this is an advanced technology vehicle that uses little or no fuel at all, but we didn't see any reason to compromise on its design.'

The only drawback at the moment is the fact that a technological breakthrough is required to make the car a reality, as a large lithium-ion battery is required to power the car! However, it shouldn't be too long before this is achieved, as some experts predict that such a battery could be production-ready by 2010–2012, which is not that far away!

Sacramento Bee-*readers speak out!*

The following are some testimonials from US citizens from an article on electric vehicles in the *Sacramento Bee* newspaper,[4] whose question of the month was:

> 'With electric vehicles so much in the news of late – including the recently unveiled Chevrolet Volt concept, designed to be propelled solely with electric power – can you envision yourself driving an all-electric vehicle in the future? And why?'

> 'You bet I'd buy an electric vehicle, if it's affordable. Why? Because I'm retired and don't drive more than 20 miles a day. The Chevy Volt would be ideal. The gas engine would hardly ever run (since it's used solely as a generator for distance beyond 40 miles). Build it, and I will buy it.'
>
> Glenn B Davis, Meadow Vista

> 'I was given the opportunity to drive General Motors' EV 1 for two weeks several years ago. The electric car is quiet and accelerates smoothly, since it has direct drive and no transmission. The EV 1 had insufficient range, lost battery strength with too much rapid acceleration (or with air conditioning on) and the rear seat was sacrificed to make room for a big battery pack. With the advent of better batteries or fuel cells and self re-charging mechanisms, these objections would be overcome. I was able to hook the EV 1 to a 110-volt house current for over-

night charges, and there were charging stands at various sites and public parking lots… . The improved all-electric car will be quiet, odorless (no gas fumes), full of pep, have plenty of room/accessories and will be both an environmental joy and a salute to freedom and foreign oil – and, in fact, all fossil fuels. It will be a win for everyone. I will want one. I would even want one if it had some sort of efficient booster engine fueled by gas, diesel, hydrogen or some other alternative fuel.'

Alvin J Livingston, Elk Grove

'Absolutely I can see myself driving an all-electric vehicle in the future, bring it on! Or should I say, bring it back! Unlike other power alternatives, the technology and infrastructure already available in the United States make the electric car a feasible alternative to fossil-fuel vehicles. Hydrogen is far off and expensive; biodiesel is promising but doesn't have the necessary infrastructure. Relying on corn-based ethanol is not an energy-saving silver bullet since industrial corn farming uses petroleum products in every step of production and delivery. I use an electric scooter, since I have a disability, and plugging it in at the end of the day is no big deal.'

Chris Weir, Sacramento

'My wife and I have been driving all-electric vehicles since 1998. First we had the GM EV 1, and now it's the Toyota RAV 4 EV. Our experience has been great. We also have a gas car for longer trips. We use the gas-fueled car once or twice a month, but we wish we didn't need to do so. The Chevrolet Volt concept is a real breakthrough; a real game-changer. It eliminates the need for the gas-fueled car. It has a range-extending engine/generator that allows a range of 600 miles or more, but runs on electric power alone for the first 40 miles or so each day. It can be charged overnight by plugging it into a standard, 110-volt electric outlet at home. We're looking forward to the

Chevrolet Volt (as GM hopes to bring the new concept to market in a few years) and cars like it. With the range problem solved, the electric car is a great way to enjoy instant, smooth power and acceleration – with no transmission, no shifting, and no pollution for almost all our driving. Widespread use would greatly reduce our dependence on petroleum.'

Tom and Vera Dowling, Folsom

'I can't imagine not driving an electric powered vehicle some day soon. Why? Because it is the cleanest transportation technology we have, and the infrastructure of electric outlets is already in place in nearly every building in America. It still takes too much energy to get usable hydrogen energy for a car, and coupled with solar and wind power, you cannot beat electricity for being cheap, clean and domestic.'

Alexandra Paul, Pacific Palisades

'Get me into the Chevrolet Volt as fast as you can. It is not a pure electric vehicle, but I want it. I can charge off my solar panels and really do something against global warming. A big plus is that I will not be subject to high gas prices, and I can help create energy security.'

Eugen Dunlap, Davis

'What a great day it will be when I get into my electric car to do my driving around town. No more gas stops, no smog, no noise. What a joy it would be. If everyone would see the 2006 documentary, *Who Killed the Electric Car?*, we would have a better understanding of why we don't have these cars available today. Put my name on the list. I want one!'

Gary Musser, Lincoln

As you can read from the above testimonies of some US citizens, electric vehicles will be sure to take off once they become available, and affordable to the public, and not just in the USA, but other parts of the world too.

Further information on the Chevrolet can be found at GM's website, <www.gm.com>, under the fuel-solutions section.

Toyota has also of course developed the Toyota Prius, which has been on the market for some time. This car manages an average (city/motorway) respectable forty-six mpg by combining a conventional petrol engine with a sixty-seven-bhp electric motor. This vehicle is now in its second-generation model.[5]

Toyota have also recently unveiled their new FT-HS concept car, which accelerates from nought to sixty mph in about four seconds, and is powered by a hybrid engine.

There is also the sporty Tesla Roadster, which is an electric vehicle assembled by Lotus in the UK for the US market. It will have a long range – some 220 miles – on one charge of its lithium ion batteries, and an amazing performance, nought to sixty mph in some four seconds! Compared to a traditional combustion engine, the Tesla's electric motor produces almost immediate acceleration and torque, with no gear changes to worry about, and all in virtual silence. According to Tesla's website, the batteries will last for about 500 charges, which equates to about 100,000 miles, with better still, hardly any servicing, as there are no moving engine parts, oils, coolants etc., which need to be changed anyway! Who said electric cars would be dull and boring? The car does come with a price tag of some US$100,000, approximately £50,000. The first cars have been rolling off the production line from October 2007.[6]

These are just a few examples of companies that are developing future hybrid high-performance, high-efficiency vehicles, which will help lower CO_2 emissions within the transport sector in the future.

In fact, electric vehicles were developed over 100 years ago. Their demise was no doubt the result of the favoured, faster internal combustion engine and the fossil-fuel frenzy brought on by the Industrial Revolution.

Electric trains have of course been around for a long time, but opting to take any sort of train is far more environmentally friendly than flying. In fact catching the train, especially in Europe, is a very pleasant, relaxing experience. A fast train can be caught in London, which whisks you under the Channel to Brussels or Paris in just over two hours. Two and a half hours later you can be in Cologne, Germany. In fact European train companies have announced recently that they will be linking up their systems, making it possible to purchase one ticket to say Budapest in Hungary, for example. At present multiple tickets would have to be bought. It is the car, however, that is responsible for the majority of transport-related greenhouse gas emissions, particularly CO_2. In view of the rapidly expanding economies of China and India and the explosion in car ownership that will bring, reducing global-warming greenhouse-gas emissions from cars will be very difficult, but essential, if pollution from the transport sector is going to be reduced.

Eurostar train service has recently made a commitment to reduce carbon emissions by twenty-five per cent by 2012. Catching the train is already ten times less polluting than flying, but not only that, Eurostar will offset any remaining emissions, so travelling on the Eurostar will be carbon neutral from 14th November 2007, once the new Eurostar service opens at St Pancras International.[7]

Aeroplanes will also become more fuel-efficient. The Airbus A380 and Boeing's Dreamliner are the most fuel-efficient yet, but they still burn conventional aircraft fuel. Richard Branson's company, Virgin, has set aside vast sums of money, about £1.6 billion (US$3 billion) to develop biofuels for transport, and to investigate and develop the technology to use biofuels as a fuel for aircraft jet engines. The technical problem of conventional biofuels freezing up when used at high altitudes will need to be overcome before this becomes possible.

So, it seems that the next generation of vehicles and more fuel-efficient means of transportation are currently in production. However, with the developed and developing worlds' increasing populations, with the ability and desire to travel and own cars, pollution from transportation is unlikely to decrease any time soon.

Here are some travel facts showing CO_2 emissions from various forms of transport.

If you happen to drive a Toyota Prius you are responsible for producing in the region of 2.1 tonnes of CO_2 a year (example based on driving 12,000 miles a year). The cost of offsetting this amount of CO_2 would be £12 or US$24.

If your vehicle of choice is a larger sports utility vehicle or 4x4, then you would be responsible for emitting some 6.65 tonnes of CO_2 into the atmosphere each year, based on the same mileage. The cost to offset would about £42 or US$84.

Every time you take a short-haul trip somewhere you are responsible for emitting 0.35 tonnes of CO_2 (example based on a two-and-a-half hour return flight). The cost of offsetting this amount of CO_2 would be £2.15 or US$4.27.

Each time you fly on a long-haul trip you are responsible for emitting 1.95 tonnes of CO_2 (based on an eight-hour return flight). The cost of offsetting these emissions would be only £11.55, US$22.98.

To find out how you can offset your emissions from flying and driving please see Chapter Y.

Figures are calculated using <www.climatecrisis.net> website using a dollar rate of $1.99 to the pound, but the exchange rate will of course vary.

Key points

- Transport accounts for approximately fourteen per cent of global greenhouse gas emissions.
- Aviation accounts for about two per cent of global emissions, but growing rapidly. Shipping accounts for almost 4.5 per cent.
- Emissions from aviation are expected to triple over the next forty-three years, faster than emissions from any other form of transport.

- ➢ Electric and other vehicles non-reliant on fossil fuels will become commonplace in the future.

---

1 Stern Review on The Economics of Climate Change, Annex 7.
2 Ibid.
3 WWF.
4 *Sacramento Bee* newspaper, <www.sacbee.com>.
5 Toyota Prius, <www.toyota.com/prius/specs.html>.
6 Tesla Motors, <www.teslamotors.com>.
7 Eurostar, <www.eurostar.com>.

Opposite above: Fourteen per cent of greenhouse gas emissions come from the burning of fossil fuels (oil) in the transport sector, cars contributing the most, with emissions from aviation the fastest-growing sector.

Opposite below: Electric vehicles are being developed that will produce zero greenhouse gas emissions, as they do not require traditional fossil fuels to power them. Ethanol is also now being used as fuel, which produces a little less CO_2 than fossil-based fuels.

OIL

F

FOSSIL FUELS

In an earlier chapter we dealt with CO_2. This chapter deals with the main culprit and cause of CO_2 emissions, fossil fuels!

The generation of electricity through burning carbon-rich coal has a greater impact on the atmosphere than any other single human activity. In 2004 the power industry created 45.3 per cent of all manmade CO_2, according to the World Resources Institute (WRI) Climate Analysis Indicators Tool (CAIT). Figures for the year 2000, taken from the Stern Review on The Economics of Climate Change, show that generating power for electricity was responsible for about one quarter of all greenhouse gas emissions worldwide.[1] If one looks at emissions of CO_2 alone (ignoring all other greenhouse gases), then for the year 2000 the figure is just under thirty-three per cent, and the latest figure from 2004, as mentioned above, shows an increase for this sector to 45.3 per cent. The latest figure, however, excludes CO_2 from land-use change, which would, no doubt, have the effect of diluting the 45.3 per cent figure down a bit if land-use-change CO_2 figures were available and included.[2]

So what are fossil fuels and how are they formed?

Fossil fuels were formed deep in the Earth's crust over millions of years by enormous pressure and heat, converting dead plant and animal matter into hydrocarbons in the form of coal, oil or natural gas, which can now be dug up or drilled for and burnt as fuel.

As these fossil fuels have formed over millions of years they are no longer considered to be part of the natural carbon cycle. They have effectively been 'locked up' within the earth's crust and rocks, etc.

In 2004 alone, some 28,000,000,000 tonnes of CO_2 were released into the atmosphere, from burning fossil fuels, which is equivalent to 800 tonnes a second![3]

Eighty-one per cent of all CO_2 emissions in developed countries come from burning fossil fuels![4]

Fossil fuels for electricity generation

Of all the fossil fuels the worst culprit is coal, which is used as the fuel for coal-fired electricity power stations in the production of electricity. Generating electricity through carbon-rich coal has a greater impact on the atmosphere than any other single human activity.

The power industry produces over twice the amount of CO_2 than is produced by the transport sector.

Basically, electricity generation hasn't changed much from the nineteenth century, with large power stations burning fossil fuels for the creation of electricity, and this model looks set to continue for some time to come.

Within the rapidly expanding economies of India and China, coal-fired electricity stations will dramatically increase. In fact China, the USA and India are the top three producers in the world.

Apart from CO_2, burning coal produces a host of other chemicals, like sulphur dioxide, which causes acid rain, nitrous oxide and other heavy metals.

What makes matters worse is that coal is the world's most widely available fossil fuel, and there is almost 200 years of coal left, if it is used at its current rate. This will be pretty academic, however, as according to the latest report out by the IPCC (see Chapter I), if the burning of fossil fuels continues unchanged, the Earth could warm by as much as 6°C (10.8°F) by 2100. *This would mean almost certain extinction of most of life on Earth!*

One has only to look at pictures from recent news stories,

showing industrial cities in these developing countries, to appreciate the pollution and clouds of smog suffusing them and blocking out the sunlight. It is believed that China has the highest number of polluted cities in the world.

It is estimated that by 2030, fifty-five per cent of all power stations built in the Asia Pacific region will be coal fired.[5] According to the Stern Review, emissions from China had been forecast to overtake those of the USA by the end of the decade; however, it appears that this may have already occurred in 2006![6]

Europe's worst polluting power stations

European countries don't get let off the hook either. According to the WWF Europe's top ten polluting power stations in terms of CO_2 output are found in the following countries:[7]

Germany (six)
Italy (one)
UK (one)
Poland (two)

These four countries produce about 203,000,000 tonnes of CO_2 every year. The UK has one power station in the list, the Drax power station, which produced 22.8 million tonnes of CO_2 in 2006. Despite producing this amount of CO_2, the power station in North Yorkshire is one of the UK's most efficient, producing about four megawatts of power.

Germany has six of the most inefficient power stations in the EU, with only Greece in front having two of the most inefficient, when comparing relative emissions (grams of CO_2 per kilowatt hour) within the EU's twenty-five countries.

Spain and Poland make up the two remaining countries in the EU having the top ten most inefficient power stations.

United States' worst polluting power stations

Texas has the most power plants in the dirtiest top-fifty list, with five plants producing a combined 85,256,075 tons of CO_2 in the year 2005.

The five plants are as follows:

- Martin Lake produces 21,593,119.5 tons of CO_2 and is ranked fifth.
- WA Parish 20,702,994 tons and ranked sixth.
- Monticello 17,491,541.6 tons and ranked thirteenth.
- Limestone 13,486,031.5 tons and ranked at thirty-two.
- Fayette Power Project 11,982,386.4 tons and ranked at forty-sixth.

As far as emission rates are concerned – that is, CO_2 pollution per megawatt-hour of electricity generated – again Texas comes out on top with six plants in the top-fifty list. The Hudson Power Plant in New Jersey tops the list, however, with an emission rate of 3,573 pounds of CO_2 for each megawatt hour of electricity generated. For the sake of comparison, the Gerald Gentlemen Plant, ranked fiftieth in the list, produces 2,383.23 pounds to produce the same amount of electricity.[8]

Please note a UK metric tonne will be the same measurement as a US metric ton, 1,000 kilograms/2,204.6 pounds. 'Tonne' comes from French/old English usage. A US ton or short ton is equal to 2,000 pounds. It is assumed all figures use metric tonne/ton measurements.

Generally a large 2.4 megawatt coal-fired power station will burn about 2,700,000 tonnes of coal a year and emit about 10,000,000 tonnes of CO_2 per annum in the process.

The additional CO_2 from all these new power stations being built in the developing world will be huge and no doubt further increase the global-warming effect as millions of tonnes of CO_2 are pumped into the atmosphere.

There is hope, however, that the newly built power stations will use new technologies that will improve the CO_2 emissions, by using carbon capture and storage technology, which will be considered in Chapter Y.

An interesting thought is that according to the WWF, if consumers in industrialised nations unplugged their telephone chargers and turned off their appliances instead of leaving them on standby, twenty-four fewer power stations would be needed. Think of all the CO_2 that could be prevented from getting into the atmosphere if everyone did his or her little bit to help?[9]

Fossil fuels for transport

The transport sector in 2000 accounted for about fourteen per cent of global greenhouse gas emissions, or twenty per cent of CO_2 (ignoring other greenhouse gases) in 2003. CO_2 emissions for 2004 remain smaller at 19.1 per cent. The majority of these emissions are from road transport, with aviation second, followed by shipping, with rail last. Emissions from the aviation sector are expected to grow the fastest, and by 2050 aviation is expected to be responsible for five per cent of the total warming effect.[10]

Fossil fuels for manufacturing and industry

This sector produced about fourteen per cent of global greenhouse gas emissions in the year 2000. Approximately ten per cent consisted of CO_2 emissions from the combustion of fossil fuels in manufacturing and construction, and three per cent were CO_2 and non-CO_2 emissions from other industrial processes. The 2004 figures for CO_2 emissions alone for the sector show an increase to 18.8 per cent, but this excludes CO_2 from land-use change and other industrial processes, so it is difficult to make a direct comparison with the 2000 figures.[11, 12]

Fossil fuels for heating offices and homes

About eight per cent of global greenhouse gas emissions come directly from heating and cooking in both commercial and residential buildings. These figures are slightly misleading, however. Really they are greater, as the latter sector are also consumers of the electricity and heat produced by the power sector.[13]

Household CO_2 emissions

Have you ever thought how much CO_2 is emitted by the average family home? Well, it's about 6.57 tonnes. This is far more than the average family car produces in one year, unless of course you drive a gas-guzzling 4x4 or SUV, that is. It would cost about £50 or US$99.5 to offset this amount of CO_2 annually when calculating the carbon offset (using the Climatecare website).[14] In the UK heating buildings accounts for about twenty-six per cent of energy demand.

Please see Chapter Y for more on carbon offsets and the tools to calculate your own carbon footprint.

So, while developed and especially developing nations continue to build power stations to gobble up the Earth's supply of fossil fuels, serious harm is being done to the atmosphere, which will continue to elevate global temperatures as higher CO_2 levels add to the greenhouse effect. The graph overleaf shows how CO_2 caused solely from burning fossil fuels has increased since the year 1900.

There are in fact enough fossil fuels left to take the world to levels of CO_2 concentrations of about 750 ppm and above[15] (about 365 ppm more than present), which would raise temperatures to such an extent that it would be catastrophic for all of Earth's inhabitants. CO_2, as a result of burning fossil fuels, is only one of the greenhouse gases that is contributing to the warming of the atmosphere. In the next chapter we will look at all of Earth's greenhouse gases, which are essential to life on Earth, but which are now warming the planet as their levels steadily increase. Manmade sources of these gases are also looked at in a bit more detail in the next chapter.

Key points

- In 2005 alone some 28,000,000,000 tonnes of CO_2 were released into the atmosphere from burning fossil fuels.
- In 2004 the generation of electricity accounted for about 45.3 per cent of all human-produced CO_2.

➢ According to the WWF, twenty-four fewer coal-fired power stations would be needed if everybody in industrialised nations turned off their mobile phone chargers and other electrical gadgets when not in use!

---

1 Stern Review on The Economics of Climate Change, Part III.
2 CAIT, <www.cait.wri.org>.
3 WWF, <www.panda.org> ('What's the cause?').
4 CAIT, <www.cait.wri.org>.
5 WWF, <www.panda.org> ('Asia's coal parasite').
6 Netherlands Environmental Assessment Agency, <www.mnp.nl>.
7 WWF, <www.panda.org> ('Dirty Thirty' May 2007 report).
8 US Environmental Integrity Project, <www.dirtykilowatts.org>.
9 WWF, <www.panda.org> ('Wasting Energy').
10 Stern Review on The Economics of Climate Change, Part III.
11 CAIT, <www.cait.wri.org>.
12 Stern Review on The Economics of Climate Change, Part III.
13 Ibid.
14 Climatecare, <www.climatecare.org.uk>.
15 Stern Review on The Economics of Climate Change, Part III.

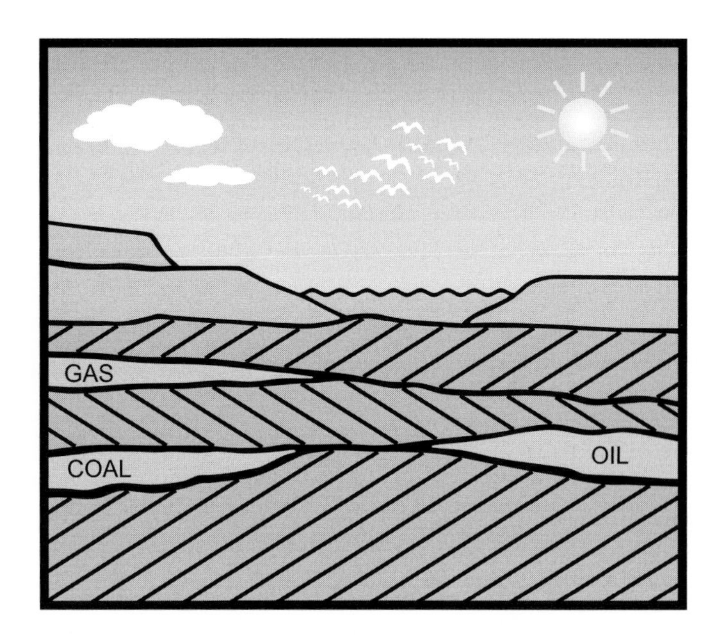

Above: Fossil fuels such as gas, coal and oil form deep in the Earth's crust over millions of years.

Below: Fossil fuels are used for energy production in coal-fired power plants, in our homes and offices, and for all our transport needs. CO_2 from burning coal for power and burning oil for transport accounted for about 64.4 per cent of CO_2 emissions in 2004.

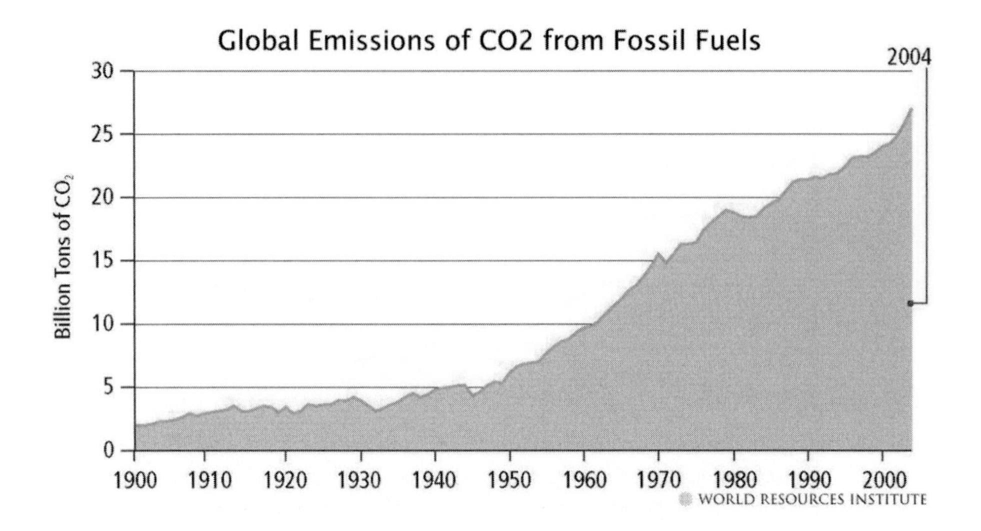

Credit CAIT World Resources Institute.

G

GREENHOUSE GASES

'Greenhouse gases' are not the gases you might find lurking about the greenhouse at the bottom of your garden! It's a term used collectively to describe the natural gases that make up the Earth's atmosphere, which trap the sun's energy, and warm up the surface of our planet.

The main gases are the following:

CO_2 (already discussed)
Water vapour
Methane
Ozone
Nitrous oxide

These are all naturally occurring gases, and without them the Earth would be far too cold to support life as we know it. In fact the 'greenhouse effect' created by these gases keeps the Earth at an average temperature of about 15°C (59°F), and without them the Earth would be a frozen ice world of about -18°C (-0.4°F).

How does the greenhouse effect work?

Well, as the sun's energy heats up the Earth's surface, after having first passed through the Earth's atmosphere, about a third of it is radiated back into space (infrared radiation). Some is then trapped or absorbed by the greenhouse gases, and about half is absorbed by the land, ocean, trees, etc. These naturally occurring gases act as a blanket, keeping in the heat, hence the 'greenhouse effect'.

So, what's the problem?

Well, everything would be okay if it were not for the manmade (anthropogenic) increase in these gases, caused in the main by the burning of fossil fuels, as previously discussed.

The main naturally occurring greenhouse gases, in the order of the effect they have on atmospheric warming, are as follows:

1. Water vapour
2. CO_2
3. Methane
4. Ozone

Water vapour

Water vapour you might think sounds pretty harmless, and on its own it probably would be! Humankind has little control over the amount of water vapour in the atmosphere, as it is a naturally occurring gas. The atmosphere contains anywhere between one to four per cent of water vapour by volume. It is technically the strongest greenhouse gas and its presence has an amplifying effect on the warming process as it acts as a positive feedback mechanism in the warming cycle. As the atmosphere gets warmer, greater evaporation occurs from the oceans and lakes on Earth. This adds to the greenhouse effect, thus warming the atmosphere further and increasing the amount of water vapour within the air. Water vapour leaves the atmosphere via precipitation (rainfall), but the warmer it gets, the more water vapour the atmosphere can hold until equilibrium is reached.[1] Evidence shows, as expected from basic physics, a warmer atmosphere holds more water, and traps more heat, amplifying the initial warming.[2]

We have discussed CO_2 already in some detail in Chapter C, so what about methane?

Methane

Methane is a gas and is found in abundance in the Earth's crust. Each molecule is about twenty times more powerful than a molecule of CO_2 in its effects as a greenhouse gas over a fixed period of time. In other words, one tonne of methane has the same effect as twenty tonnes of CO_2. Fortunately there is far less of it present in the atmosphere, and it stays in the atmosphere for about only ten years, compared with hundreds of years for CO_2. Nevertheless, levels of methane have increased since pre-industrial times and it is a very potent greenhouse gas.[3]

The amount of methane in the atmosphere amounts to about 1,775 ppb (parts per billion), which is more than double the level it was in the 1750s, prior to the Industrial Revolution. According to the NASA Earth observatory methane levels have stayed flat for the past seven years (as of 2006), after having risen for the previous twenty years. Almost two-thirds of methane emissions can be linked to human activities, and scientists believe that the slowdown in methane concentration may be linked to leak-preventing repairs that have been made to oil and gas pipelines and storage facilities, among other things. According to Professor Sherwood Rowland,

> 'If one really tightens emissions, the amount of methane in the atmosphere in 10 years from now could be less than today. We will gain some ground on global warming if methane is not as large a contributor in the future as it has been in the past century.'

Professor Rowland was the co-recipient of the 1995 Nobel Prize for discovering that chlorofluorocarbons in products such as aerosol sprays and coolants were damaging the Earth's protective ozone (see below).[4]

Believe it or not, large amounts of methane are produced by animals, particularly cows and other livestock, through burping and flatulence!

Methane and melting permafrost

By far the most worrying prospect is the release of large deposits of methane from melting permafrost (permanently frozen topsoil) in Arctic regions, like Siberia, and from deep under the sea. It is believed that large deposits of methane lie deep under the ocean in the form of frozen hydrate deposits. As these two locations become affected by global warming there is a danger that this methane could be released back into the atmosphere, which would have serious consequences for Earth's climate as it reacts with oxygen and produces CO_2.

Scientists have been studying the permafrost in Siberia and have found that its thawing and the warming and drying of wetland areas could release methane and CO_2 into the atmosphere in the future. One scientific model suggests that up to ninety per cent of the upper layer of permafrost will thaw by 2100.

These stores represent more than double the total cumulative emissions from fossil fuel burning so far!

It is estimated that in northern Siberia methane emissions have already increased by sixty per cent since the mid-1970s.[5]

Methane trapped as gas hydrates under the ocean, it is suggested, may also be affected if ocean warming penetrates deeply enough to destabilise these trapped pockets of the gas. Immense quantities may be trapped under the ocean, equivalent to twice as much CO_2 as is present in all coal, gas and oil reserves remaining. The gas hydrates are kept in place under the oceans in very cold regions and under enormous pressure, which keeps them stable. If these hydrates were released, possibly from warming of the Earth's oceans, then the methane stores could be released into the atmosphere, causing a significant contribution to greenhouse gases and therefore increased global warming.[6]

What about ozone?

Ozone (O_3) is a molecule that is made up of three atoms of oxygen, and amounts to a tiny fraction of the atmosphere. Ozone was first discovered by the German chemist Christian Friedrich Schönbein, in 1840. Ozone, however, plays a very important role as it protects Earth's inhabitants from harmful ultraviolet radiation, and without it we would all be much more susceptible to skin cancer, cataracts and impaired immune systems.[7]

Ozone resides mainly in the stratosphere, which is the region of atmosphere between ten and forty kilometres (about six and twenty-five miles) above the Earth. It is here where the ozone acts as a shield to protect us from the radiation.

Ozone can also be found in the troposphere, which is the layer from Earth's surface to about ten kilometres above (about six miles). The ozone found here is a harmful pollutant and forms petrochemical smog when nitrogen oxides and hydrocarbons mix to form ozone and other ingredients.

Ozone has been measured in the atmosphere since the 1920s by ground-based instruments, where calculations of its concentration could be made from the atmosphere above the location where the instruments were situated. Satellites however have now given scientists the ability to look at ozone levels on a daily basis over the entire Earth.

The USA has a programme run jointly with NASA and NOAA, which has been observing and measuring ozone changes over the last twenty years.[8]

Most of us have been aware of the hole in the ozone layer for some time now. This is a separate issue to that of global warming, and scientists now know that stratospheric ozone is being depleted worldwide, partly as a result of human activities. The large polar ozone losses, particularly over Antarctica, are as a direct result of effects of human-produced chemicals, in particular, chlorofluorocarbons (CFCs).[9]

Ozone levels drop so much during the Antarctic spring that scientists describe the loss as the 'Antarctic ozone hole'.

CFCs were commonly used in fridges, and air-conditioning units, but their use has been phased out by the signing of the Montreal Protocol Treaty in 1987, which limited their production and usage as scientists realised what they might be doing to the ozone layer. A revised treaty led to their complete phase-out in 1996, and as a result most CFC concentrations have started to decrease around the Earth.

The production and use of less harmful CFC replacement

hydrochlorofluorocarbons (HCFCs) are also scheduled to be phased out by about 2030.

As a result, scientists believe that the Antarctic ozone hole will repair itself in about fifty years or so.

Scientists are continuing to learn more about ozone and the processes that affect it, as well as mankind's contribution to the hole in Earth's protective ozone layer. NASA will help with their new Earth observing system satellites (EOS), which carry a sophisticated set of instruments that will measure the interactions within the atmosphere that affect Earth's protective ozone shield.

What about nitrous oxide?

Nitrous oxide is given off naturally by bacteria in the oceans and soils. Mankind adds to the levels of this gas by using it in fertilisers for agriculture, livestock farming, in aerosols, and by burning fossil fuels in car engines. It is also used as a propellant in rocket engines. The gas is more commonly known as 'laughing gas', but it's no joke to think that each molecule of it is 296 times more powerful than CO_2, in consideration of the potential for global warming over a fixed period of time. It also hangs about in the atmosphere for about 120 years! Nature, however, is responsible for far more nitrous-oxide emissions than mankind, but levels have risen since pre-industrial times as a result of reasons given above.

What do these gases have in common?

The characteristic that all these gases have in common is their ability to 'force' the Earth's climate. The IPCC (see Chapter I) defines climate forcing as

> 'An externally imposed perturbation in the radiative energy budget of the Earth climate system, e.g. through changes in solar radiation, changes in Earth albedo, or changes in atmospheric gases and aerosol particles.'[10]

According to the NOAA, climate forcing is described as

> 'Radiative forcing of climate by trace gases is commonly referred to as the "greenhouse effect". Solar radiation that passes through clouds and that is not reflected back into space strikes the Earth's surface. The longer wavelength (infrared) radiation created there is reflected upwards, and then is absorbed by clouds and the greenhouse gases (greenhouse gases include CO2, methane, nitrous oxide, etc.). Those constituents reradiate upwards and downwards, thereby heating the Earth's surface.'[11]

The IPCC takes the pre-industrial era, arbitrarily chosen as 1750, as its baseline, and NOAA research shows that all the major greenhouse gases (water vapour excluded, as it is a feedback rather than direct forcing mechanism) have had a positive radiative forcing (warming) on the climate since 1750.

The greenhouse effect is often talked about in terms of the effect the equivalent concentration of CO_2 in parts per million has on the atmosphere and includes the effects of the six Kyoto greenhouse gases. Under the Kyoto Protocol, which will be discussed further in Chapter K, six greenhouse gases were recognised and classified with the aim of reducing their emissions. The gases in question are CO_2, methane, nitrous oxide, sulphur hexafluoride, HFCs and PFCs.

The warming effect caused by all Kyoto greenhouse gases emitted by mankind is now equivalent to about 430 ppm of CO_2. This can also be expressed as 430 ppm CO_2e. These levels are higher now than at any time in at least the past 650,000 years.[12]

In other words the warming effect of all the greenhouse gases increases the levels of CO_2 (385 ppm) by about 45 ppm equivalent.

Even if greenhouse-gas levels were stabilised at today's level (430 ppm) global mean temperatures would eventually rise to about 1 to 3°C (1.8–5.4°F) above pre-industrial levels, that's 2°C (3.6°F) more than at present.[13]

The problem is that greenhouse gas levels are unlikely to be stabilised and will continue to rise. If they continue at their current level, concentrations would be more than treble pre-industrial levels by 2100, committing the world to a 3 to 10°C (5.4–18°F) warming based on the latest climatic projections.[14]

'A warming of 5°C (9°F) on a global scale would be far outside the experience of human civilisation and comparable to the difference between temperatures during the last Ice Age and today.'[15]

Where do manmade greenhouse gases come from?

The main causes of worldwide greenhouse gas emissions in descending order based on the most recent figures (available from year 2000) are as follows:

1 Power sector, twenty-four per cent

Most of the greenhouse-gas emissions come from the generation of power and heat for both domestic and commercial buildings, and from emissions generated by the transformation of fossil fuels into a form of fuel that can be used in transport, buildings and industry.

2 Land use, eighteen per cent

This is mainly CO_2 emissions from deforestation as discussed in Chapter D.

3 Transport/agriculture/industry, fourteen per cent each

Transport is responsible for the third largest source of greenhouse gas emissions jointly with emissions from agriculture and industry. Three quarters of emissions from transport are from road transport, with aviation producing about one eighth of those emissions and rail and shipping the remainder. Agriculture is also responsible for fourteen per cent of greenhouse gas emissions with virtually all of them consisting of greenhouse gases other than CO_2, methane and nitrogen for example. Industry accounts for fourteen per cent of emissions, most of which consists of CO_2, mainly from the burning of fossil fuels.

4 Buildings/waste/other energy, sixteen per cent

Buildings account for eight per cent of the remainder of greenhouse gas emissions, with waste and other energy accounting for the rest. The direct burning of fossil fuels for heating and cooking in both residential and commercial premises accounts for these emissions, however the overall figure is probably higher due to the fact that this sector is also a consumer of electricity and heat provided by the power sector (as mentioned above).[16]

World greenhouse gas and CO_2 emission levels

To give you some idea of the amounts of greenhouse gases being emitted by these sectors annually, figures from the WRI, using their CAIT, give the following emissions figures for the year 2000 for greenhouse gases and CO_2 separately:[17]

1. Greenhouse gas emissions (six Kyoto gases, CO_2 / methane / nitrous oxide / perfluorocarbons / hydroflurocarbons / sulphur hexafluoride) from the above sectors amounted to about 43,000 million tonnes of CO_2 equivalent, or 43 gigatonnes.

2. CO_2 emissions alone, i.e. not including the other gases mentioned above from the same sectors, amount to about 33,000 million tonnes of the above figure.

CO_2 therefore accounts for about seventy-five per cent of greenhouse gases produced by these sectors in 2000. These figures have almost certainly increased since the year 2000!

CO_2 and Greenhouse gas emissions per country

Data from year 2004 figures from CAIT (most recent CO_2 emission figures) show that the USA was in first place as the worst polluter, emitting some 5,888,000,000 tons of CO_2 (5,888 $MtCO_2$), which equated to 19.8 per cent of world CO_2. China followed with 5,204

$MtCO_2$ (17.5 per cent), then the EU's twenty-nine countries with 4,412 $MtCO_2$ (14.84 per cent). Total greenhouse gas emissions for the latest year from which data is available (2000) show the USA in the lead with 6,867 $MtCO_2$ (19.15 per cent of world total), with the EU twenty-nine countries 5,312 (14.81 per cent), and China emitting 4,882 (13.61 per cent). Thereafter, the table is as follows:

4th,	Russia
5th,	India
6th,	Japan
7th,	Germany
8th,	Brazil
9th,	Canada
10th,	UK

For greenhouse gas emissions including land-use change, after the USA, the EU twenty-nine and China, the ranking is as follows:

4th,	Indonesia
5th,	Brazil
6th,	Russian Federation
7th,	India
8th,	Japan
9th,	Germany
10th,	Malaysia

The UK is ranked thirteenth on this basis.

Each person in the USA is estimated to be responsible for about 20.1 tons of CO_2, compared with about 4 per person living in China (China has a far larger population). Data from 2007 however reveals China may have just recently overtaken the USA as the worst polluter. Between 1990 and 2002 Chinese emissions grew by forty-nine per cent compared with eighteen per cent for the USA over the same period.

Generally, as expected, emissions from developing nations are growing the fastest.

While all these activities are adding millions of tonnes of greenhouse gases (primarily CO_2) into the atmosphere, manmade greenhouse gases are still a very small proportion compared to natural emissions caused by decaying vegetation, plant and animal respiration, etc. Manmade greenhouse gases have however started to upset the natural balance, it would seem, between the amount nature puts into the atmosphere and takes out through natural carbon sinks such as the forests, oceans and land. This is evidenced by the fact that levels of these gases, methane, CO_2 etc., are now higher than they have been for hundreds of thousands of years, as evidenced by direct measurements and from ice-core data.

The consequences of increased greenhouse gas levels will be an increase in temperature, which, even if it increases by a relatively small amount, could be serious, and this will be looked at further in Chapter T.

We will now look back in time to see how climate has changed over Earth's history in order to put global warming into some kind of perspective, in the next chapter, Historical Climate Change.

Key points

- Earth's greenhouse gases keep the planet at a comfortable 15°C, without which Earth would have a temperature of about -18°C.
- Anthropogenic (manmade) additions to these gases are amplifying the greenhouse effect and causing global warming.
- Methane is a very potent greenhouse gas. Methane gas from frozen methane hydrates under the sea and under the permafrost, if released, could cause runaway global warming.
- Ozone acts as a shield and protects the Earth from harmful radiation. The hole in the ozone layer has been partly caused by the use of CFCs, which have now been phased out.
- Global greenhouse gas levels (including CO_2) are now about 430 ppm. Global CO_2 levels are now 385 ppm.

- Electricity generation produces about twenty-four per cent of global greenhouse gas emissions, or if looking at CO_2 alone, forty-five per cent of global emissions in 2004.

---

1 NOAA, <www.ncdc.noaa.gov> (on greenhouse gases and water vapour).
2 Stern Review on The Economics of Climate Change, Part I.
3 NASA, <www.earthobservatory.nasa.gov> (on methane, impacts on climate change, 18th-July 2005).
4 NASA, <www.earthobservatory.nasa.gov> ('seven-years stability of methane may slow global warming', 26th November 2006).
5 Stern Review on The Economics of Climate Change, Part I.
6 Ibid.
7 NASA, <www.nasa.gov> (on ozone).
8 NASA, <www.earthobservatory.nasa.gov>.
9 Ibid.
10 IPCC Climate Change 2001, Working Group 1 Scientific Basis.
11 NOAA, <www.esrl.noaa.gov>.
12 Stern Review on The Economics of Climate Change, Part I.
13 Ibid.
14 Ibid.
15 Ibid.
16 Op cit, Part III.
17 CAIT, <www.cait.wri.org>.

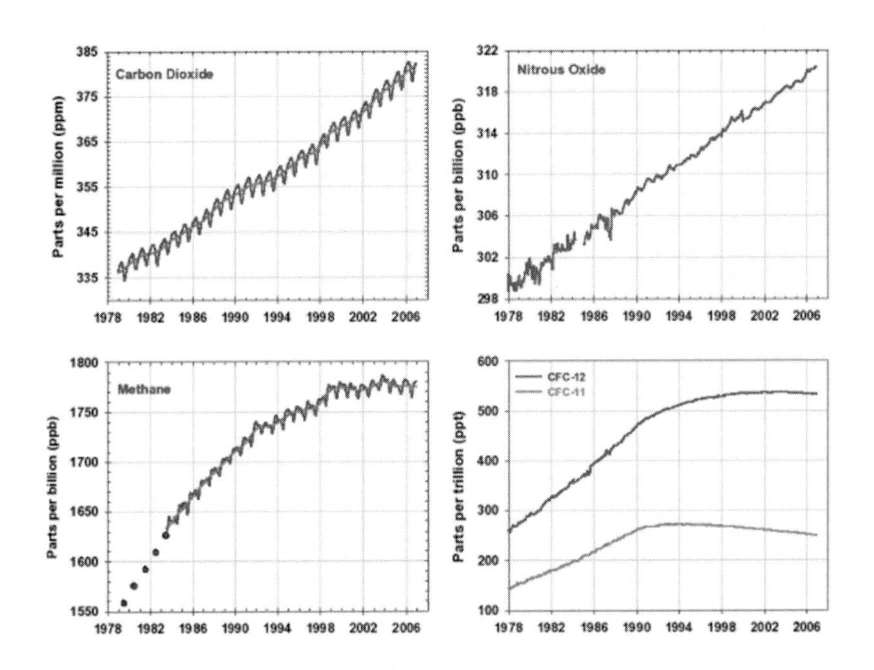

Global averages of the concentrations of the major, well-mixed, long-lived greenhouse gases – CO_2, methane, nitrous oxide, CFC-12 and CFC-11 from the NOAA global flask sampling network since 1978. These gases account for about ninety-seven per cent of the direct radiative forcing by long-lived greenhouse gases since 1750.

Credit NOAA, <www.esrl.noaa.gov>.

Opposite above: 1 Energy from the sun warms the Earth's surface.
2 Earth is cooled as infrared radiation escapes into space.
3 Greenhouse gases trap infrared radiation warming Earth's atmosphere.
4 The greenhouse effect keeps the Earth at a comfortable average temperature of plus 15°C.

Opposite below: 1 Higher levels of greenhouse gases, particularly CO_2*, increase Earth's greenhouse effect and cause further warming of Earth's atmosphere, as more of the sun's energy is trapped.*

2 As Earth's atmosphere warms up, temperatures start to rise. Permafrost regions are already melting and releasing methane into the atmosphere. It is feared that frozen methane hydrates stored under the oceans could warm and release their stores of methane gas, causing significant further warming. Greenhouse gas levels now stand at 430 ppm CO_2 *equivalent, higher than at any time than in the last 650,000 years.*

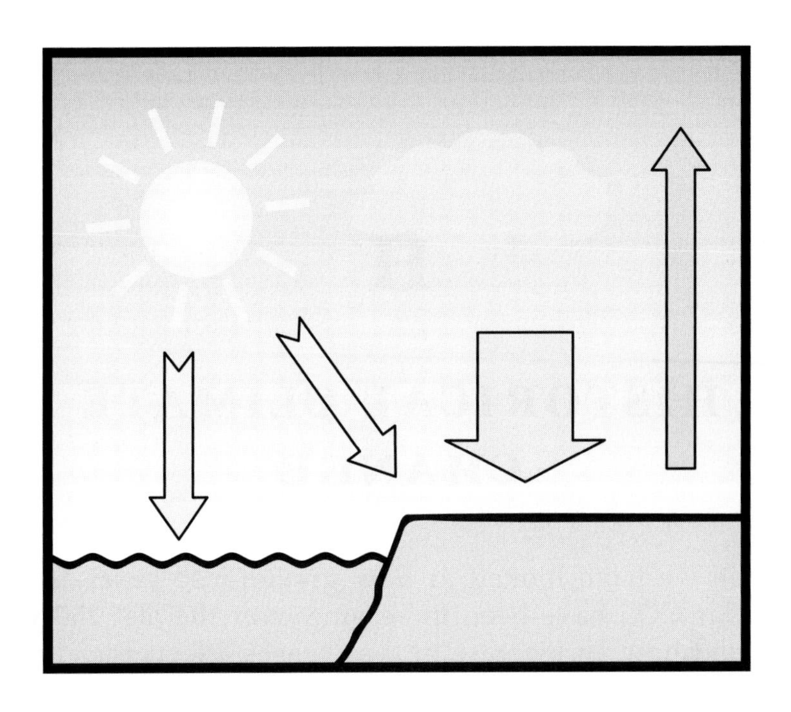

H

HISTORICAL CLIMATE CHANGE

So far in we have looked at how greenhouse gases, and in particular CO_2, have been increasing over the last 250 years or so, and how an increase in these gases affects the climate. In Chapter T we will look in more detail at how increasing levels of greenhouse gases affect Earth's temperature and will reveal that in Earth's recent history temperatures have been increasing.

When talking about recent history we mean very recent on a geological scale, i.e. in the last hundred years or so, which is a minuscule period in Earth's history, bearing in mind it's some 4.5 billion years old!

Indeed, we have learnt from Chapter C that accurate measurements of CO_2 have been recorded by Charles Keeling since 1958, from the Mauna Loa observatory in Hawaii, and that CO_2 levels have risen from 315 ppm to 385 ppm during the last fifty years. It also seems that an increase in CO_2 levels is linked with an increase in temperature.

Just a natural variation?

Science has enabled us to look into Earth's distant climatic past, which may just give us an insight as to what we can expect in the future.

The study of past climate, or palaeoclimatology to give it its scientific name, reveals that a huge variation in climate has taken place over millions of years of Earth history.

When we think about the distant past, of the days when dinosaurs were roaming the Earth, the vision that is usually conjured up is of a hot and balmy climate. Indeed there is evidence that the Earth's oceans 65,000,000 years ago were about 10 to 15°C (18 to 27°F) warmer than today. We also know that the Earth has, in the past, been much colder, cold enough to plunge it into the grip of many an Ice Age.[1]

During the last fifty years or so, scientists have made huge advances in unlocking Earth's remote climate by looking at data and evidence preserved in tree rings, sea corals, and pollens (biological data). Then there are rock and other sediments from sea and lakebeds (geological data), and the most revealing is frozen ice-core data from the Antarctic and Greenland ice sheets (cryological data).

We know that the Earth is about 4.5 billion years old by using a process called radiometric dating. Mass spectrometry is used to date the element uranium, which decays at an incredibly slow but predictable rate. Carbon is also used to age-date material (carbon dating), but can only be used to date relatively young material – about 50,000 years old. These methods allow scientists to tell roughly how old something is, and this technique is used to date a variety of data/materials mentioned above.

Chilling data

Ice cores have proved to be invaluable in revealing what the Earth's climate has been like in the past. Scientists have managed to extract cylindrical cores of ice, typically about four inches in diameter, and extending to over three kilometres deep into the Antarctic and Greenland ice sheets.

How is this done? As snow forms, it crystallises around minute particles in the atmosphere. Microscopic bubbles of gas get trapped as layer upon layer of snow falls to the ground, which can then be analysed to reveal what the temperature, CO_2 and methane levels were at any given time in the past.

The Vostok ice cores taken from Antarctica records climatic data going back about 400,000 years, but the deepest ice cores, recently extracted from the Antarctic by a European project (EPICA), have pulled out the deepest core allowing a glimpse back some 900,000 years in time.[2]

What do ice core records reveal about Earth's climate?

Well, if we take a journey back through Earth's climatic history, we find that between about 1350 and 1850 there were two distinct cold periods, referred to as the 'Little Ice Age', when temperatures were about 1°C (1.8°F) lower than today's.[3] Rivers and lakes in the northern hemisphere would regularly freeze over during winter, and indeed the Thames in London, over a period of about 400 years, would freeze over on average every twenty-five years or so. The freezing of the river may have been assisted, however, by the slower-moving water due to old London Bridge in situ at the time.

If we go back further, from about 1000 AD to about 1300, we arrive at a period known as the 'Medieval Warm Period'. Temperatures during this time were warmer than they had been for thousands of years before that, but similar to now.[4]

Back further still, from between 10,000 and 15,000 years ago, we arrive at the end of the last Ice Age or glacial period. The period from then, which we are still in, is called the Holocene Interglacial Period, and marks a time when global temperatures started to increase as the grip of the last Ice Age ended. We are reminded that we are indeed in an interglacial period, which suggests at some point in the future there is likely to be another ice age, unless of course global warming either prevents or delays this from happening. Chapter M deals with the possible causes of ice ages in more detail later on.[5]

The last Ice Age lasted for about 100,000 years, and during this period massive ice sheets up to 2 kilometres deep (1.24 miles) extended from the North Pole as far down as New York! Sea levels as a result of all the ice fell by about 100 metres (328 feet) below current levels.[6] Antarctic ice-core data reveals that during this period, about 20,000 years ago, temperatures were almost 8–10°C (about 14–18°F) lower than present.

As we go back further still, to about 115,000 years ago, the records show that temperatures started to rise again. Indeed this indicates that there was another interglacial period from about 115,000 to 140,000 years ago. There was a corresponding increase in temperatures by almost 8–10°C (14–18°F) during this period, bringing temperatures up to a level not much different from what they are today.

Further back again, yes, you guessed it, we enter another ice age. From about 140,000 years ago to about 240,000 years ago the Earth was in the grip of another 100,000-year-long ice age, with a corresponding drop in temperature by almost 8–10°C (14–18°F) for most of that period.

The ice records reveal that this cycle has repeated itself at least eight times, and sediment cores show evidence of ice ages occurring on a cyclical basis over the last million years, every 10,000 to 15,000 years or so, and lasting for about 100,000 years each time. There were ice ages lasting a relatively shorter 40,000 years or so for millions of years prior to that. The timings of the ice ages may be linked to the Earth's orbit around the sun, and this is looked at further in Chapter M.[7]

So, we know that the Earth, during the last 650,000 years at least, has been gripped by several 100,000-year-long ice ages, with 10,000 to 15,000-year-long interglacial warm periods, when temperatures fluctuated by about 10°C or thereabouts (18°F). While the Antarctic ice-core records show temperature variations of +/ - 10°C (18°F) or thereabouts between glacial/interglacial periods, mean global temperatures would probably have varied by about fifty per cent of that of the poles, i.e. by about 5 to 6°C (9 to 10.8°F).

It is clear therefore that Earth's climate has cycled from ice age to interglacial over the millennia, and during our current interglacial period there has been some small, one to two degree plus or minus variation in temperature that has resulted in the Medieval Warm Period and the Little Ice Age. In fact the Earth has experienced a series of ice ages during the last 2.6 million years or so.[8]

CO_2 levels and ice-core records

Well, this is the interesting bit, as scientists have discovered that CO_2 levels are now higher than at any time in the past 650,000 years.[9] It is also the case that when the data relating to CO_2 is compared with past temperatures, it can be seen that with CO_2 levels at their highest, so are temperatures, and vice versa, and these measures in turn correspond to Earth's past glacial and interglacial periods.

The lower the level of CO_2, the lower the temperature of Earth is in a glacial period. As CO_2 levels increase, so does the temperature, which corresponds to interglacial periods in Earth's history.

CO_2 levels are now about 385 ppmv in air, and increasing compared to a 650,000-year range of between 180 and 300 ppmv, as revealed by the ice-core records. It doesn't take a genius to work out that this can only mean one thing based on past data – higher temperatures!

Does CO_2 lead temperature or lag behind it?

While higher levels of CO_2 go hand-in-hand with higher temperatures, does this mean that higher CO_2 levels have caused the rise in temperature, or could it be the other way round? Some scientists dispute the theory that higher CO_2 levels are causing higher temperatures, as there appears to be approximately an 800-year time lag between CO_2 rises and temperature increase. While the ice-core data cannot be questioned, it seems the data may be open to interpretation, meaning that higher temperatures cause higher levels of CO_2 rather than the other way round. Ice-core records show CO_2 changes over very long timescales, that is glacial to interglacial time periods, and while CO_2 levels and temperatures do appear to be inextricably linked, the science community now think that temperatures may be the first to rise during interglacial periods (less ice equals less reflection of sunlight, which equals warmer temperatures). Warmer temperatures mean higher greenhouse gas concentrations. CO_2 and ice volume should therefore lag behind temperature somewhat when looking at glacial to interglacial timescales.

Global average temperature was lower during glacial periods for two main reasons:

1. CO_2 levels were only about 190 ppm in the atmosphere, and other greenhouse gases were lower.
2. The Earth's surface was a lot more reflective because of the presence of much more sea and surface ice, giving a much greater albedo.

It is thought that the second factor has the greater influence, creating two-thirds of the total radiative forcing, with CO_2 and other greenhouse gases the other third. Therefore while temperature is certain to rise based on the current levels of CO_2 (385 ppmv), the second factor (above) needs to be taken into account when extrapolating temperature against ice-core CO_2 records. In other words, CO_2 and other greenhouse gases may be responsible for only just over thirty per cent of the radiative forcing or warming found in the ice-core records.[10]

So what if it gets a bit warmer, it's better than another ice age, right?

Well, apart from the Earth's ice sheets melting (see Chapter N), historical proxy data reveals an event that occurred in Earth's distant past, about 55,000,000 years ago. The event caused global mass extinction, some 10,000,000 years after the killer asteroid that ended the dinosaurs' reign on Earth.

Palaeocene-Eocene Thermal Maximum

The Palaeocene-Eocene Thermal Maximum (PEMT), as it is known, saw a sudden global spike in air and sea temperatures over a period of only a few thousand years. It is thought that ocean temperatures went up by about 5 to 8°C (9 to 14.4°F) during this period. Two scientists from the Scripps Research Institute in California found, from looking at deep-sea sediment cores, that an underwater conveyer-belt-like process, in which cold and salty water exchanges with warmer surface water, virtually shut down in the southern hemisphere, and started up in the northern hemisphere. This apparently drove warmer water to the deeper sea, possibly releasing the previously frozen methane gas hydrates discussed in the last chapter. This would have caused a sudden massive spike in greenhouse gases, which would have warmed the Earth further, resulting in a mass extinction of bottom-dwelling marine life and mass migrations on land as animals adjusted to the new climate.[11]

Younger Dryas period

A more worrying and slightly more recent rapid climate-change event occurred about 14,500 years ago, towards the end of the last Ice Age, called the Younger Dryas period. The period is named after an Arctic-alpine plant, Dryas, which populated Europe during these cold conditions. The Earth's climate warmed fairly rapidly about 14,500 years ago, only to change suddenly again to conditions more akin to the Ice Age that had just ended. This glacial spell lasted for about 1,000 years. Then, about 11,500 years ago, temperatures rose by about 10°C (18°F) in a decade or so, incredibly quickly. This sudden change in climate from relatively warmer to ice-age conditions is thought to have been triggered by the breakdown of the ocean water conveyor system, which brings warmer water and air to the northern hemisphere from the tropics. This, in turn, is thought to have been caused by a meltwater surge from Antarctica or North America, as ice sheets began to melt, resulting in huge amounts of freshwater flowing into the ocean. This has been termed Meltwater pulse 1A following a

> '...defined sea level rise of about 16-24 metres (52-79 feet) about this time. A further surge of meltwater took place after the Younger Dryas period known as Meltwater Pulse 1B, when meltwater from glacial Lake Agassiz, southwest of Hudson Bay in Canada drained into the North Atlantic.'[12, 13]

The ocean thermohaline conveyor still plays an incredibly important role in maintaining northern-hemisphere temperatures, and will be looked at further in Chapter O.

If these rapid climate-changing events can occur in the Earth's distant past, what's stopping something similar from happening again if the Earth's oceans warm up and glaciers and ice shelves start to melt as a result of increased greenhouse-gas warming?

A sliding-scale graph of ice-core data going back in time 400,000 years, from the Vostok Antarctica core, can be seen on NASA's Earth observatory website.[14]

In the following chapter we will look at the IPCC, which is the world's leading scientific body on climate change, and which sets out the latest scientific consensus on global warming and climate change.

Key points

- The Earth is about 4.5 billion years old.
- The Earth has been through many ice ages, and is currently in an interglacial stage, called the Holocene.
- CO_2 levels are now higher than at any time in the last 650,000 years, as confirmed by ice-core records.

---

1 NOAA, <www.ncdc.noaa.gov> ('The fire and ice before').
2 European Ice Project for Ice Drilling in Antarctica (EPICA), <www.esf.org>.
3 NOAA, <www.ncdc.noaa.gov>.
4 NOAA, <www.ncdc.noaa.gov>.
5 NOAA, <www.ncdc.noaa.gov> (a paleo perspective on global warming).
6 Ibid.
7 Ibid.
8 Ibid.

9 Stern Review on The Economics of Climate Change, Part I.
10 Real Climate, <www.realclimate.org> (the lag between temperature and CO_2).
11 Mongabay, <www.mongabay.com>.
12 NASA, <www.giss.nasa.gov> (Vivian Gornitz, January 2007).
13 NOAA, <www.ncdc.noaa.gov>.
14 Earth Observatory, <www.earthobservatory.nasa.gov/study/paleoclimatology-icecores/>.

I

INTERGOVERNMENTAL PANEL ON CLIMATE CHANGE

The Intergovernmental Panel on Climate Change (IPCC) was established in 1988 by the World Meteorological Organisation and the United Nations Environmental Program. Its purpose was to assess the scientific, technical and socio-economic information relevant to the understanding of climate change.

The panel evaluates research and studies that have been carried out by many thousands of researchers, together with many hundreds of scientists from around the world. The information is then collated by a number of IPCC working groups and summarised and released in the form of assessment reports, each dealing with different aspects, ranging from the science of climate change, the impacts of climate change and what can be done to mitigate the problem. Each report also has a summary for policymakers as well. The first report was released in 1990, the second in 1995, and further reports appear every six years thereafter. The latest report is the fourth assessment report (AR4 released its summary for policymakers on 2nd February 2007).

The reports have become the main reference literature on the subject of global climate change, and the findings in the reports have been cited and quoted whenever the subject of climate change is discussed.

Each report from 2001 can be found in its entirety on the internet at <www.ipcc.ch>. So, for the purposes of this chapter, a few salient points have been taken from each of the reports from 1995 to 2007, which give a clear picture of how the panel have interpreted the scientific data and research that has been collated on global climate since 1995.

1995 report

From the 1995 'Summary for Policymakers, Second Assessment Report', the following observations were made:

1. 'The atmospheric concentrations of greenhouse gases, inter alia, carbon dioxide (CO_2), methane (CH_4), nitrous oxide (N_2O) have grown significantly: by about 30 per cent, 145 per cent and 15 per cent respectively (values for 1992). These increases can be attributed largely to human activities, mostly fossil fuel use, land use change and agriculture.'
2. 'Global mean surface air temperature has increased by between about 0.3 and 0.6°C [0.54–1.26°F] since the late 19th Century. Additional data available since 1990 and the reanalyses since then have not significantly changed this raise estimated increase.'
3. 'Global sea level has risen between 10 and 25 cm [3.9–9.8 inches] over the last 100 years and much of the rise maybe because of increase in global mean temperature.'[1]

2001 report

From the 2001 'Summary for Policymakers, Third Assessment Report' the following observations were made:

1. 'Global average surface temperature has increased over the 20th century by about 0.6°C [1.08°F].'
2. 'Global average sea level has risen and ocean heat content has increased. Tide gauge data show that global average sea level rose between 0.1 and 0.2 metres during the 20th century [3.9–7.8 inches].'
3. 'Snow cover and ice extent have decreased. There has

been widespread retreat of mountain glaciers in non-polar regions during the 20th century.'[2]

2007 report

From the 2007 'Summary for Policymakers, Fourth Assessment Report' the following observations have been made:

1. 'The updated 100 year linear trend (1906–2005) of 0.74 (0.56 to 0.92) degrees Celsius (1.33°F) is therefore larger than the corresponding trend for 1901–2000 given the TAR of 0.6 (0.4 to 0.8) degrees Celsius [1.08°F].'
2. 'Carbon dioxide is the most important anthropogenic gas. The global atmospheric concentration of carbon dioxide has increased from a pre-industrial value of about 280 ppm to 379 ppm in 2005. The atmospheric concentration of carbon dioxide in 2005 exceeds by far the natural range over the last 650,000 years (180–300 ppm) as determined from ice cores.'
3. 'Warming of the climate system is unequivocal, as is now evident from observations of increases in global average air and ocean temperatures, widespread melting of snow and ice, and rising global average sea level.'
4. 'Most of the observed increase in globally averaged temperatures since the mid-20th century is very likely due to the observed increase in anthropogenic greenhouse gas concentrations.'
5. 'Mountain glaciers and snow cover have declined on average in both hemispheres. Widespread decreases in glaciers and ice caps have contributed to sea level rise' [ice caps to not include contributions from Greenland and Antarctic ice sheets].
6. 'Observations since 1961 show that the average temperature of the global ocean has increased to depths of at least 3,000 metres [9,842 feet] and the ocean has been absorbing more than 80 per cent of the heat added to the climate system. Such warming

causes sea-water to expand, contributing to sea level rise.'

7 'New data (since the TAR) now show that losses from the ice sheets of Greenland and Antarctica have very likely contributed to sea level rise over 1993 to 2003.'
8 'Global average sea level rose at an average rate of 1.8 (1.3 to 2.3) mm [0.07 inches] per year over 1961 to 2003. There is high confidence that the rate of observed sea level rise increased from the 19th to the 20th century.'
9 'Temperatures at the top of the permafrost layer have generally increased since the 1980s in the Arctic (by up to 3 degrees C) [5.4°F].'
10 'Changes in solar irradiance since 1750 are estimated to cause a radiative forcing of +0.12 (+0.6 to +0.30) Wm-2, which is less than half the estimate given in the TAR.'
11 'The annual carbon dioxide concentration growth rate was larger during the last 10 years (1995–2005 average: 1.9 ppm per year), than it has been since the beginning of continuous direct atmospheric measurements (1960–2005 average: 1.4 ppm per year) although there is year-to-year variability in growth rates.'[3]

Strengthening of opinion

It can be seen that when the reports are looked at over the decades, they show a clear strengthening of scientific views on the subject. It can be seen that atmospheric greenhouse gases have increased, and CO_2 levels are now higher than at any time during the last 650,000 years.

The panel concludes that it is very likely that most of the observed temperature increases since 1950 are due to manmade (anthropogenic) greenhouse gas concentrations.

The Earth's average temperature was initially assessed as increasing by between 0.3 and 0.6°C over a hundred-year period.

The panel later concluded that temperatures have increased by about 0.74°C (1.33°F) over a similar timescale.

The Earth's glaciers and ice caps are melting, and ice sheets from Greenland and Antarctica have very likely contributed to the sea-level rise observed between 1993 and 2003, which has been seen to have been increasing since 1961 at an average rate of 1.8 millimetres per year (0.07 inches).

All in all the latest IPPC report provides for pretty sobering reading. The terms 'very likely' and 'high confidence' used in the reports can be interpreted as meaning a ninety per cent chance of an outcome or result and an eight-in-ten chance of being correct, respectively!

Prize winners!

In October 2007 the Norwegian Nobel committee recognised the work of the IPCC and Al Gore jointly for their efforts in disseminating knowledge on climate change, by awarding both of them the 2007 Nobel Peace Prize.

The author is responsible for Fahrenheit and imperial conversions from IPCC, measurements.

In Chapter J we look at a very surprising connection to global warming, jet trails.

Key points

- ➢ The IPCC is considered the world's leading scientific panel on climate change, and has released four reports over twenty-seven years specifically dealing with climate change and global warming.
- ➢ The latest report confirms that the Earth has warmed by 0.74°C (1.33°F) between 1906 and 2005.
- ➢ Temperature increase since the 1950s is due to manmade greenhouse-gas concentrations.
- ➢ CO_2 is the most important anthropogenic gas.
- ➢ Warming of the climate system is unequivocal.

- Global sea levels are rising and temperature has increased by up to 3°C (5.4°F) at the top of the Arctic's permafrost layer.

---

1 IPCC, Science of Climate Change 1995, Working Group 1.
2 IPCC, Science of Climate Change 2001, The Scientific Basis.
3 IPCC, Science of Climate Change 2007, The Physical Basis.

J

JET TRAILS

Technically called contrails or condensation trails, the white condensation trails left behind by flying aircraft are the result of hot humid exhaust from the aircraft's jet engines mixing with the surrounding air of lower humidity and temperature.

Possible connection with global warming?

Following the terrible attacks on 11th September 2001 on the USA, scientists were given the opportunity to study the effects of the contrails, for the first time in decades. The skies above the continental USA were contrail-free for three days while all aircraft were either grounded or diverted from US airspace.

According to NASA, satellite observations during air-traffic shutdown enabled their scientists to gain insight into the atmospheric conditions that govern the formation of contrails – clouds caused by aircraft emissions. Patrick Minnis, a senior research scientist at NASA's Langley Research Center, said:

> 'Because air traffic is expected to grow over the next 50 years, contrail coverage will also increase and may significantly impact the Earth's radiation budget by 2050.'

The Earth's radiation budget, the balance between the planet's incoming sunlight and outgoing heat energy, drives climatic change. It is thought that contrails can spread into extensive high, thin cirrus clouds that tend to warm the Earth because they reflect less sunlight back into space than the amount of heat they trap.

Contrails typically form in large numbers from overlapping commercial flights, making it difficult for scientists to follow their development. The air-traffic shutdown gave Minnis and his team the chance to track individual persistent contrails from military aircraft on 12th September.

According to Minnis,

> 'Six aircraft were responsible for the formation of Cirrus clouds that covered more than 20,000 square kilometres within an area between Virginia and Central Pennsylvania. During normal days, the area is crossed by thousands of jetliners that could each produce contrails similar to those from military jets.'[1]

Cloud connection?

NASA's scientists it seems have found that cirrus clouds formed by contrails from aircraft engine exhaust are capable of increasing average surface temperatures enough to count for a warming trend in the USA that occurred between 1975 and 1994. Cirrus clouds exert a warming influence on the surface by allowing most of the sun's rays to pass through, but then trapping some of the resulting heat emitted by the surface and lower atmosphere.

So, believe it or not, the white trails left by jet aircraft as they streak across the skies may well have an effect on global warming, though it would seem that studies into the link are in their infancy. No doubt NASA scientists will be looking into the connection further as air travel increases together with the temperature of the planet! In the UK British Airways (BA) are also researching the possible connection between jet trails and global warming. A recent news report confirmed that BA will be gathering data from their aircraft's emissions to try and assess the effect of aircraft emissions at altitude.

In the next chapter we will look at what has been done politically to try and solve the global warming problem when we look at the Kyoto Protocol.

Key points

- ➢ Jet trails, or condensation trails, are the hot humid exhaust from an aircraft's jet engine, which appear white after mixing with air of relative lower humidity and temperature.
- ➢ Jet trails can cause cirrus clouds to form, which can in turn trap the sun's heat and cause a warming trend by increasing Earth's surface temperature.

---

1 NASA, <www.earthobservatory.nasa.gov/newsroom/im ages> (aircraft contrails).

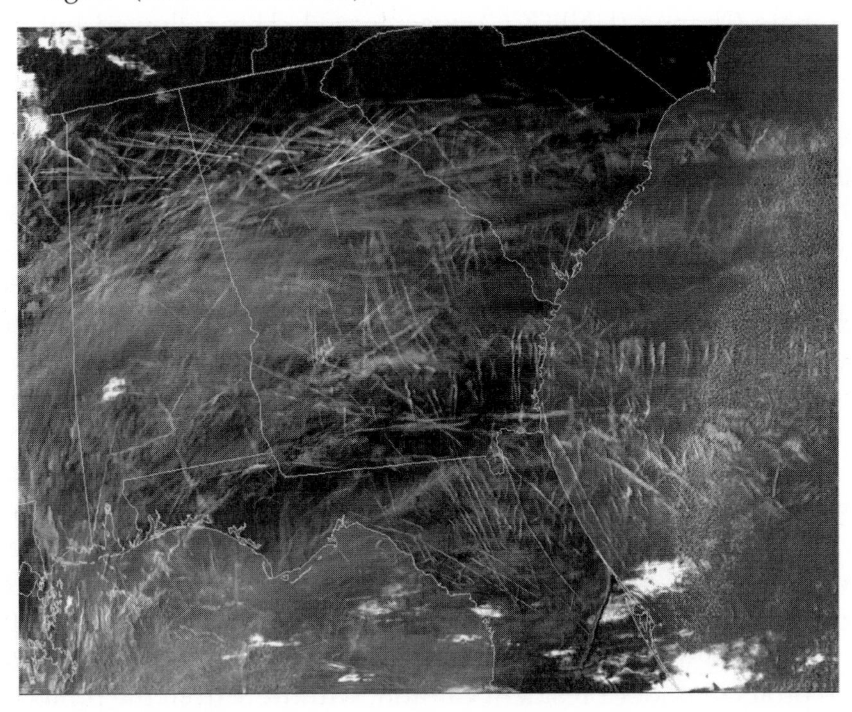

In 2004, NASA scientists discovered that contrail-generated cirrus clouds could be responsible for much of the warming of surface temperatures over the USA from 1975 to 1994.

Credit NASA earth observatory, <www.earthobservatory.nasa.gov>.

K

KYOTO PROTOCOL

The major political force that brought global climate change to the fore was born of the Rio Earth Summit in 1992, where an agreement called the United Nations Framework Convention on Climate Change (UNFCCC), was agreed. This followed hot on the heels of the release of the IPCC's first report on climate change.

The Kyoto Protocol, as it became known, entered into force on 16th February 2005, and became the first important step as governments and countries around the world committed themselves to a binding agreement to reduce their greenhouse gas emissions.

The road to Kyoto was a bumpy one. After agreeing the UNFCCC, governments realised that action had to be taken to set real reduction targets, and so, in 1997, in Kyoto in Japan the parties to the UNFCCC reached agreement on what later became known as the Kyoto Protocol.

For the protocol to enter into force it had to become ratified by at least fifty-five parties to the convention, and incorporating a list of thirty-five industrialised nations. These nations together with the EU (collectively termed Annex 1 countries) accounted for at least fifty-five per cent of the total of CO_2 emissions in 1990.

A stalling point emerged when the USA, having signed up to the protocol under President Clinton, withdrew from the protocol when President Bush was elected in early 2000. Luckily, the protocol was given a lifeline by Russia, when President Putin ratified the agreement on 18th November 2004.[1]

So what is the aim of the Kyoto Protocol?

The protocol requires industrialised countries to reduce their emissions of greenhouse gases by five per cent below 1990 levels by 2008–2012.[2] The gases covered are the six main greenhouse gases, namely

Carbon dioxide (CO_2);
Methane (CH_4)
Nitrous oxide (N_2O)
Hydrofluorocarbons (HFCs);
Perfluorocarbons (PFCs)
Sulphur hexafluoride (SF_6)

How does the protocol work?

To put it in simple terms, the agreement works by providing for various market-based mechanisms to assist countries or individual companies meet their respective emission targets. Emissions caps are put on Annex 1 countries, giving each country an emissions quota, or allowable amount of CO_2 emissions. So for example the EU has been given a target to cut greenhouse gas emissions by eight per cent below 1990 levels. This target can be distributed among member states.

In general, developed Annex 1 countries have to reduce their CO_2 emissions. Developing non-Annex 1 countries, however, have not had to cap their emissions, but instead will participate in CO_2 emission-reducing projects.

> 'The Kyoto Protocol provides for detailed reporting and accounting for emissions and emissions allowance allocations within annex 1. Prior to each commitment period over which emissions reductions will be made, parties are required to submit initial reports establishing their "Assigned Amount", the emissions a country will be expected to emit over that period. If they exceed this they will have to purchase credits (allowances) from others that have emitted less than their assigned amount.

> Annex 1 countries must submit detailed annual emissions data on an annual basis in national inventory reports, with supplementary information on allowance holdings and transactions. Failure to submit annual reports and inaccuracy in reports can lead to suspension of eligibility to participate in the Kyoto mechanisms.'[3]

There are three basic methods open to countries to meet their targets:

1 Emissions trading schemes
2 Clean development mechanisms
3 Joint implementation projects

Emissions trading schemes (ETS)

As explained above, these schemes allow for the trading of CO_2 emissions for carbon credits. So, if a country or industry exceeds its assigned amount of CO_2 emissions, it would be able to purchase credits from a country or industry that has not. Only a small proportion of global emissions are covered by these schemes, and currently the EU has the largest scheme, the EU ETS.

Clean development mechanisms (CDM)

This is a way for Annex 1 countries to earn credits by investing and funding climate-friendly projects and technologies in developing countries, thus helping control emissions in these countries.

Joint implementations projects (JIP)

Basically these are the same as CDMs, but with Annex 1 countries investing in climate-friendly technology in other Annex 1 countries, rather than other developing countries.

In reality each country will pass its emissions quota on to individual greenhouse-gas-emitting companies or industrial sectors, with the ability to purchase credits using the trading schemes with other countries/companies, or earn credits from investing in CDM/JIP projects. The basic principle behind Kyoto is that developed countries actively cap their emissions as they were responsible for most of the greenhouse gases present in the atmosphere from the industrialisation process, while investing billions of pounds into climate-friendly technologies for developing countries, climate studies and research.

The expansion of such carbon-trading markets across the Kyoto member states will have the effect of spreading carbon-friendly technology via the CDM and JIP projects, with the overall aim of reducing CO_2 emissions in the process.

> 'The Protocol to its credit has established an aspiration to create a single global carbon price and implement equitable approaches to sharing the burden of action on climate change.'[4]

While Kyoto is an incredible achievement it is at present the world's only agreement attempting to limit greenhouse gas emissions and global warming. There are problems due to the fact that the USA has not ratified the protocol, and neither had Australia, until literally 3rd December 2007, following a change of government. They were the only two developed nations not to have done so. The fact that, in April 2007, Australia faced one of its worst droughts for a long time, no doubt assisted in the decision to ratify the protocol, as soon as the political climate in the country changed! Due to the severity of the drought, it is being suggested that Australia could be one of the first developed countries to feel the effects of global warming.

> 'The protocol has of course been criticised for creating quantitative obligations only for the rich countries, without placing any constraints on emissions from the growing emerging economies.'[5]

While the USA refuses to sign up to the protocol, other countries such as India and China, while ratifying it, do not have obligations at present to reduce their greenhouse gas emissions, on the basis that these countries were not responsible for today's greenhouse gas levels. However, at the rate these countries are developing, they will soon be the world's major polluters as they build more and more fossil-fuelled electricity plants to satisfy their energy needs. Indeed, it is believed that in June 2007 China overtook the USA as the world's highest CO_2 emitters.

It is true to say, however, that instead of the richest countries reducing their emissions by five per cent to 1990 levels, they have in fact increased them by some ten per cent. It is believed that only four countries, the UK, France, Germany and Sweden are on track to meet the targets set.

Nevertheless,

> 'The Kyoto Protocol can be seen as a first stepping-stone on the path to international co-operation on climate change, given political, economic and scientific realities.'[6]

Certainly in the USA,

> 'California is currently developing specific proposals for a cap-and-trade scheme as part of its goal to reduce emissions by 25 per cent by 2020.'[7]

Recently the UK announced the introduction of a climate-change bill, making it the first country to set legally binding targets to reduce its CO_2 emissions. The bill should receive royal assent by autumn 2008, and it will set targets to reduce CO_2 emissions by sixty per cent by 2050. There are already murmurs that this may be increased to an eighty per cent reduction.

What will happen after 2012?

The UK has taken the lead here by making a commitment to reduce its CO_2 emissions after the first Kyoto commitment period comes to an end. The parties to Kyoto also met in Montreal in

2005 to discuss the second period after 2012 to develop an action plan up to 2017.

UN secretary Ban Ki-moon convened a high-level event that took place in New York on 24th September 2007, to promote discussions on ways to move the international community towards negotiations on a new global agreement on climate change. This took place at the UN climate-change conference in Bali, on 3rd December 2008.

The purpose of discussions will be to try and get in place a multilateral framework for action on climate change for the period after the Kyoto agreement ends in 2012.[8]

As at the time of writing, it seems that at least a climate road map of agreement has been reached, which the USA has accepted. However it's not clear what agreement has been reached with regard to the most important points, greenhouse-gas-emissions reductions, though a roadmap that includes measures to prevent deforestation has been agreed..

According to the WWF,

> 'We need to cut greenhouse gases by about 60–70 per cent globally by mid-century in order to stay well below a 2 degree Celsius [3.6°F] temperature increase above pre-industrial temperatures and in order to avoid the most catastrophic consequences for humans and ecosystems.'

While the Kyoto Protocol was a big step in the right direction, it seems that much more needs to be done and far greater cuts made to CO_2 emissions to ensure global temperatures do not rise over critical levels.

To find out in more detail what are considered to be critical levels, please see Chapter T.

We will now look at how global warming is affecting the world's oceans, and specifically how ocean levels have started to rise in Chapter L.

Key points

- ➢ The Kyoto Protocol came into force in 2005 and will expire in 2012.
- ➢ The aim of the protocol was for industrialised countries to reduce their greenhouse gases by five per cent below 1990 levels by 2012 – however emissions have in fact increased.
- ➢ Currently the USA remains the only developed country not to have ratified the agreement.
- ➢ The WWF advocates cutting greenhouse gases by about 60–70 per cent globally by 2050, to stay below a 2°Cs increase above pre-industrial levels.

---

1 WWF, <www.panda.org>.
2 WWF, <www.wwf.org.uk>.
3 Stern Review on The Economics of Climate Change, Part VI.
4 Ibid.
5 Ibid.
6 Ibid.
7 Ibid.
8 WWF, <www.panda.org> (Kyoto Protocol).

L

LEVEL OF THE OCEANS

As we discussed in Chapter I, the scientific consensus is that ocean levels are rising across the globe.

According to the IPCC's 'Third Assessment Report' (TAR), sea levels are predicted to rise by 9 to 88 centimetres (3.54 to 34.64 inches) by 2100.[1]

According to the NASA Earth observatory, data from multiple sources have revealed that 'The sea level rose, on average 3 millimetres (0.1 inches) per year between 1993 and 2005'.[2]

That's an increase of 3.6 centimetres (1.41 inches) over twelve years! This compares to an average rate of 7.46 centimetres (2.94 inches) over forty-two years (1961–2003) measured by the IPCC, which suggests that the rate of sea-level rise is in fact increasing. A rise in sea level towards the mid-range of the figures predicted by the IPCC would therefore seem to be more accurate, unless sea-level rise accelerates further, which would place the rise by 2100 at the top end of IPCC figures.

What causes the oceans to rise?

There are three major factors to the rise in ocean levels. Oceans rise as they become warmer (thermal expansion). Warmer water takes up more space than cooler water. Secondly, water is added to the oceans as glaciers and ice shelves melt. A third contribution comes from the reduction in salinity (salt content) in the oceans as more freshwater is added from the melting ice. Freshwater is less dense than seawater, so the same amount of freshwater takes up more space than a corresponding amount of saltwater, which means the melting ice effectively has a double impact on ocean levels.[3]

How are ocean levels measured?

Scientists have been using a combination of satellite data from the US/French satellite TOPEX/Poseidon, and more recently the replacement Jason-1 satellite together with flotation devices to measure the level of the world's oceans.

To measure sea levels, oceanographers at NASA's Jet Propulsion Laboratory relied on satellite measurements of sea-surface height (which increases as sea temperature increases), taken by TOPEX/Poseidon and later by Jason-1.

Complementing the Jason-1 satellite data were temperature and salinity measurements from the Argo float programme. By using measurements from a variety of sources, oceanographers can form a clearer picture of the oceans' behaviour in different parts of the world.

Another string to NASA's bow comes in the form of GRACE (Gravity Recovery and Climate Experiment), which can precisely measure surface height not only of the world's oceans, but also the giant bodies of ice that feed them. If ice mass height drops and ocean levels rise, GRACE can measure both changes simultaneously.

GRACE observations determined that from 2002 to 2005, Antarctic ice lost enough mass to raise global sea levels by 1.5 millimetres (0.05 inches).[4]

Researchers attribute about half of that increase to melting ice and the other half to thermal expansion as the ocean absorbs excess energy.

It is undeniable therefore that global ocean levels are on the rise. The cause, it would seem, can only be the world's oceans warming up and additional freshwater from melting glaciers and polar ice adding to sea levels, which in turn also decreases ocean salinity. Melting sea ice (as opposed to continental ice and glaciers), as we shall see in the next chapter, doesn't necessarily contribute to sea-level increase, much the same as melting ice in a glass will not cause the contents of that glass to overflow.

We know that the oceans act as a major sink for CO_2 and it has been estimated by NASA jet propulsion scientists that

> 'Over the past 40 years the oceans have absorbed 84 per cent of the excess heat – enough heat to warm the entire atmosphere by 27 degrees centigrade [48.6°F].'[5]

It seems that as a consequence of increased greenhouse gases, therefore, ocean levels are now rising.

> 'However because warming only penetrates the oceans very slowly, sea levels will continue to rise substantially more over several centuries. On past emissions alone, the world has built up a substantial commitment to sea level rise.'[6]

Earth's threatened cities

Warming from the last century has already committed the world to rising seas for many centuries to come. Further warming this century will increase this commitment. Coastal areas are among the most densely populated areas in the world, and currently more than 200,000,000 people live in coastal floodplains around the world, with 2,000,000 square kilometres (772,200 square miles) of land and a trillion dollars' worth (£500 billion) of assets less than one metre elevation (about three and a quarter feet) above sea level. Many of the world's major cities (twenty-two of the top fifty) are at risk of flooding from coastal surges. These include

1. Tokyo
2. Shanghai
3. Hong Kong
4. Mumbai
5. Calcutta
6. Karachi
7. Buenos Aires
8. St Petersburg

9 New York
10 Miami
11 London

In almost each case, the cities rely on costly flood defences for protection. Some estimates suggest that 150–200,000,000 people may become permanently displaced by the middle of the century due to rising sea levels, more frequent floods, and more intense droughts.[7]

Lost islands?

Many small island nations such as the Maldives in the Indian Ocean, or Tuvalu in the Pacific, may disappear. The low-lying Carteret Islands in the South Pacific are among the smallest remote islands in the world. They have a population of about 2,500 and their island homes may become submerged beneath the Pacific by 2015, theirs the first peoples displaced by climate change.

The Maldives, where seventy-five per cent of the land is less than one metre (about three and a quarter feet) above sea level, may be gone within thirty years according to some scientists.[8]

A host of other islands in the Pacific are doomed to follow if ocean levels keep rising as predicted.

New York under threat?

According to NASA, New York will also be under threat of severe flooding by the mid-twenty-first century. A recent study by Rosenzweig and Gornitz in 2005 and 2006, using the GISS Atmosphere-Ocean Model global climate model for the IPCC, projects a sea-level rise of 15 to 19 inches (38.1 to 48.2 centimetres) by the 2050s in New York City. Adding as little as 1.5 feet (0.45 metres) of sea-level rise by the 2050s to the surge of a category-three hurricane on a worst-case track would cause extensive flooding in many parts of the city. Areas potentially under water include the Rockaways, Coney Island, much of southern Brooklyn and Queens, portions of Long Island City, Astoria, Flushing

Meadows-Corona Park, lower Manhattan, and eastern Staten Island from Great Kills Harbor North to the Verrazano Bridge.

Of course it wouldn't just be New York under threat from rising oceans. Other great cities such as Beijing, San Francisco Bay, Florida and Venice, in addition to those above, would also be threatened by rising oceans.[9]

To read more about how glaciers and ice sheets are responding to global warming and ocean-level rise, please see Chapter N.

Having discussed water and oceans we now look at frozen water and the most widely accepted theory for the cause of the great ice ages and the sun's involvement in past climatic change – Milankovitch Cycles.

Key points

- ➢ NASA has recorded a sea-level rise of 3 millimetres a year between 1993 and 2005, or 3.6 centimetres over twelve years (1.41 inches).
- ➢ Sea levels rise because seawater is getting warmer, and warmer water expands, meltwater from glaciers and ice are adding to sea levels, and freshwater is less dense than saltwater and therefore takes up more space.
- ➢ The Carteret Islands in the Pacific may be the first islands to be lost to rising sea levels.

---

1 Stern Review on The Economics of Climate Change, Part I/ IPCC TAR.
2 NASA, <www.earthobservatory.nasa.gov>.
3 Ibid.
4 Ibid.
5 NASA Jet Propulsion Laboratory, <www.sealevel.jpl.nasa.gov>.
6 Stern Review on The Economics of Climate Change, Part I.
7 Ibid.
8 WWF, <www.panda.org>.
9 NASA, <www.nasa.gov> (on hurricane risks to New York City).

Opposite above: Jason and Grace satellites take precise measurements of ocean surface levels. Argo floats measure ocean temperatures down to a level of 2,000 metres below the surface.

Opposite below: 1 Sea warms up causing thermal expansion.

2 Melting land glaciers and continental ice sheets add to ocean level rise.

3 Less dense fresh water takes up more space than an equal mass of seawater.

M

MILANKOVITCH CYCLES

Have you ever wondered what causes the Earth's seasonal variations in climate, or whether there is any relationship between the Earth's journey around the sun and Earth's ice ages?

Well, Milutin Milankovitch (1879–1958) did! He was a Serbian mathematician and engineer and he formulated a comprehensive mathematical model that calculated longitudinal differences in insolation (incoming solar radiation), and the corresponding surface temperature for 600,000 years prior to the year 1800. He then tried to correlate these changes with the cycles of glaciation that have occurred in Earth's past.[1]

Milankovitch studied Earth's orbital variations around the sun and theorised that the waxing and waning of Earth's ice ages, or periods of glaciation, are closely linked to Earth's orbit around the sun. His theory is also supported by evidence found in deep-sea sediment cores.

As the Earth moves around the sun, its journey is governed by three separate motions:

1 Eccentricity – Earth's orbit around the sun.
2 Obliquity – tilt of Earth's axis.
3 Precession – rotation of Earth's axis.[2]

Eccentricity

We all know that the Earth orbits the sun. What Milankovitch found out, however, was that the Earth's orbit changes over time, from a slight to an exaggerated elliptical orbit, over a 90–100,000-year time period. The closet approach Earth makes to the sun on its current, slightly elliptical orbit, is called the perihelion, which occurs on 3rd January. Earth is farthest away at aphelion, which occurs on 4th July. This means that the Earth is closest to the sun when it is summertime in Australia (in the southern hemisphere), and winter in London (in the northern hemisphere), and farthest away from the sun when it's summertime in the northern hemisphere!

This helps make the northern hemisphere's winter a little less extreme, since the warming effect occurs in that season when the Earth is closest to the sun.

When the Earth's orbit is highly elliptical the incoming solar radiation at Earth's closest approach to the sun would be about twenty to thirty per cent more than at its farthest distance, providing for a totally different climate from the one at present.

The real reason for the Earth's seasons, however, is that the Earth is also tilted on its axis.

Obliquity and Earth's seasons

The Earth rotates around its axis as it orbits the sun, and is tilted from its central plane at an angle that ranges between 22.1 and 24.5 degrees. Currently the axis is tilted at 23.5 degrees from its plane of orbit as it circumnavigates the sun. The tilt also has a cycle, of about 40,000 years, moving from 22.1 degrees through to 24.5 degrees and back. At present the Earth is moving towards the upright, so that the tilt is reducing towards a 22.1-degree angle.

The tilt of the Earth's axis is what actually gives Earth its seasons. When the northern hemisphere is tilted towards the sun, during summer, the sun is higher in the sky, allowing more of the sun's energy to strike the Earth and warm it up. When the northern hemisphere is tilted away from the sun, during winter, the sun is lower in the sky, meaning less of the sun's energy hits the surface, making it cooler. During spring and autumn (fall), the Earth is neither tilted towards nor away from the sun, allowing the sun's energy to concentrate on the equatorial regions. It is of course the opposite for the southern hemisphere, as explained earlier.

This should not be confused with the fact that the Earth is farthest away from the sun, by about 5,000,000 miles, on its current elliptical orbit, during the northern hemisphere's summer (July). The Earth is actually warmer during this time, despite receiving less incoming solar radiation, purely because the Earth is tilted towards the sun, and the majority of Earth's landmass is found above the equator. The reason for this is that land heats up quicker than water, which means, ironically, the Earth is warmer when it is at its farthest distance away from the sun!

Apart from defining the seasons, the Earth's tilt means that when the tilt is greater, there is greater seasonal contrast, meaning warmer summers and colder winters in both hemispheres.

The relevance of this to periods of glaciation is, it is thought, that when the Earth's tilt is less – i.e. towards the 22.1-degree angle – cooler summers result. This allows the previous winter's snow and ice to accumulate in higher latitudes, eventually building up into massive ice sheets, giving rise to Earth's next glacial period.

Precession

This is the change in orientation of Earth's axis, sometimes described as its wobble. Not only does the Earth's axis tilt as explained above, but the axis itself has a wobble, a bit like a spinning top as it slows down. This has a cycle of about 26,000 years, and it effectively changes the orientation of the Earth with respect to its closest and farthest orbits around the sun. At present northern hemisphere summers occur in July, when Earth is farthest away from the sun, because the Earth is tilted towards the sun at this time. Over a 26,000-year timescale, northern hemisphere summers will shift to the point where they occur in January, as is the case for southern hemisphere now.[3]

As can be seen, the Earth's journey around the sun is highly complex, consisting of a trio of motions with varying timescales governing each cycle. It is these complexities that dictate Earth's climate and seasons over vast periods of time, and that look to be responsible for causing the waxing and waning of Earth's ice ages.

By studying these changes scientists may get some insight into what might happen to Earth's climate if global warming through anthropogenic (manmade) means interferes too much with Earth's natural balance.

We now move to the next chapter, 'No More Ice!', which looks at the disturbing fact that the world's ice sheets and glaciers seem to be melting and retreating at alarming rates.

Key points

- ➢ Milankovitch theorised that the Earth's motion around the sun is linked to periods of glaciation on the Earth.
- ➢ Earth's axis is tilted at 23.5 degrees on its plane of orbit and it is this tilt that gives Earth its seasons.
- ➢ Earth's journey around the sun is governed by three separate motions, which have a bearing on Earth's climate and periods of glaciation over periods of thousands of years.

---

1 NASA, <www.earthobservatory.nasa.gov> ('On the shoulders of giants').

2 Ibid.

3 Ibid.

Opposite above: The Earth's orbit changes from elliptical (high eccentricity) to nearly circular (low eccentricity) over a cycle that takes 90,000 to 100,000 years. When Earth's orbit is highly elliptical, the amount of insolation received by Earth at its perihelion (closet approach) would be twenty to thirty per cent greater than at its aphelion (furthest departure), resulting in a far different climate from what Earth experiences today.

Credit NASA earth observatory, <www.earthobservatory.nasa.gov>.

Opposite below: The Earth's axis is inclined at an angle of 23.5 degrees from the plane of its orbit around the sun. This tilt and Earth's motion around the sun cause the change of the seasons. In January, the northern half of Earth tilts away from the sun. Sunlight is spread thinly over the northern half of Earth, and the north experiences winter. At the same time, the sunlight falls intensely on the southern half of Earth, which has summer. Earth is seen in this position on the right in the above illustration. By July, Earth has moved to the opposite side of the sun, and the situation is reversed.

Credit NASA, <www.nasa.gov>.

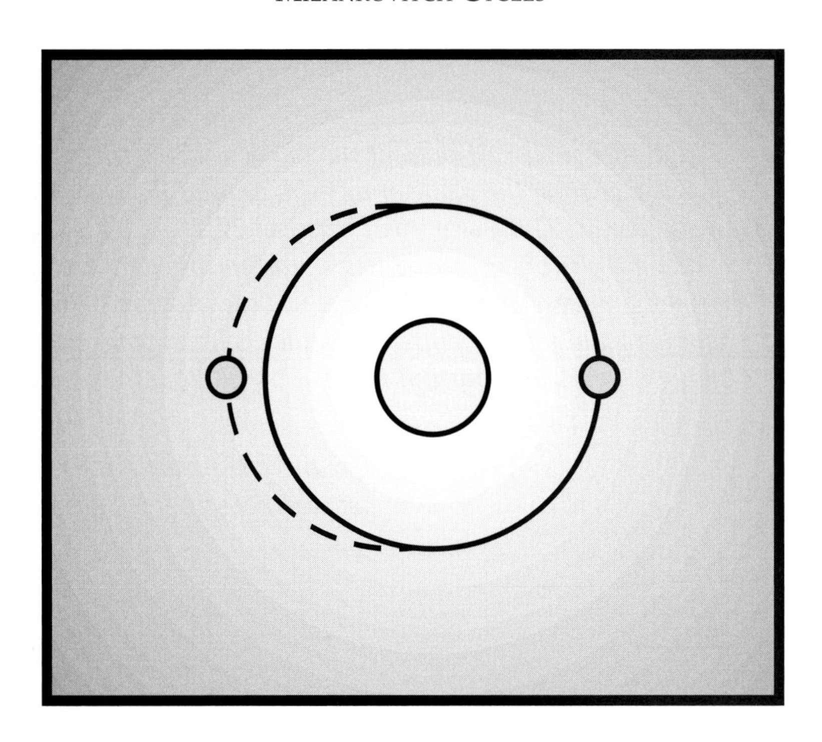

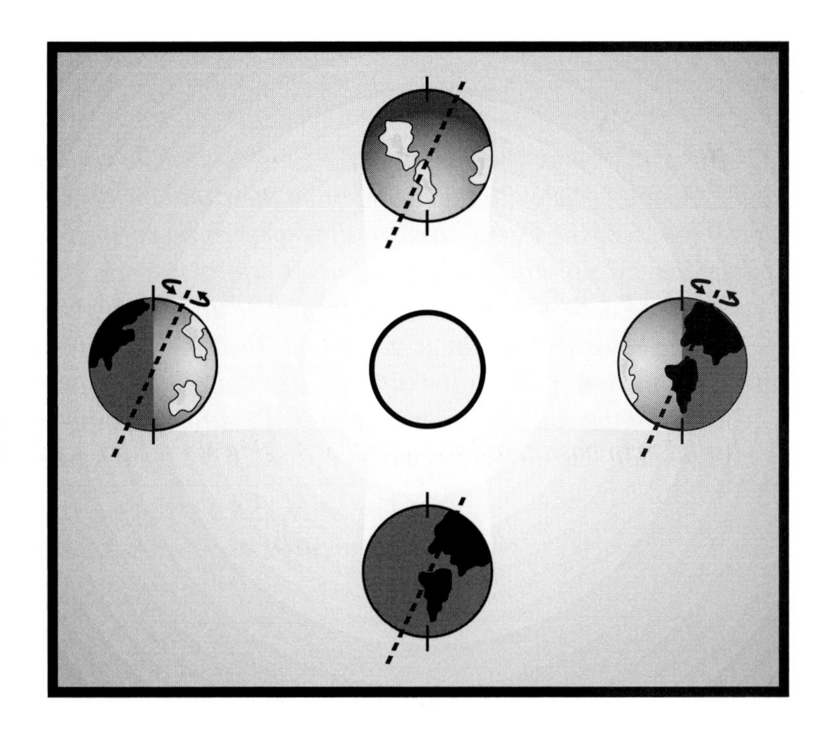

Opposite above: At present the Earth's axis is tilted at 23.5 degrees from the plane of its orbit around the sun. The tilt changes over a 40,000-year cycle between 22.1 and 24.5 degrees. When the tilt is greater Earth's seasons are more severe, i.e. warmer summers and colder winters. Less tilt results in cooler summers and milder winters. It's the cooler summers that are thought to allow snow and ice to accumulate into massive ice sheets over time.

Credit NASA earth observatory, <www.earthobservatory.nasa.gov>.

Opposite below: Precession is the Earth's 'wobble' around its axis. Over a cycle of 26,000 years or so there will be a change in orientation of the Earth's axis with respect to its perihelion and aphelion. At present northern summer occurs near its aphelion (when the Earth is furthest away from the sun). If a hemisphere is pointing towards the sun at perihelion, then it will be pointing away at aphelion, making the difference in seasons more extreme. Currently the southern hemisphere is tilted towards the sun at perihelion, but this will change over a 26,000 year timescale.

Credit NASA Earth Observatory, <www.earthobservatory.nasa.gov>.

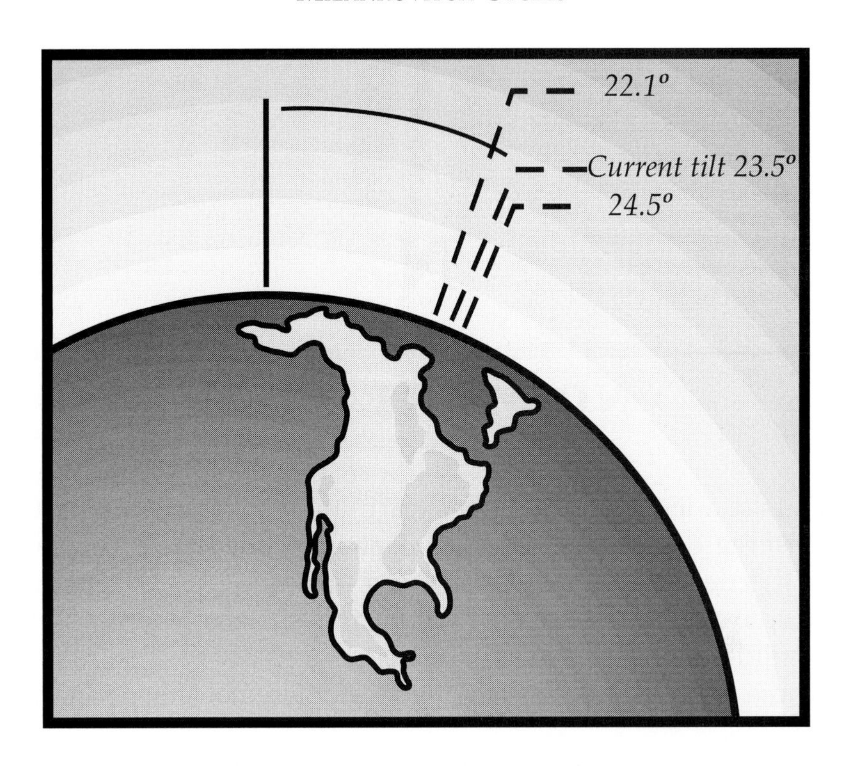
22.1°
Current tilt 23.5°
24.5°

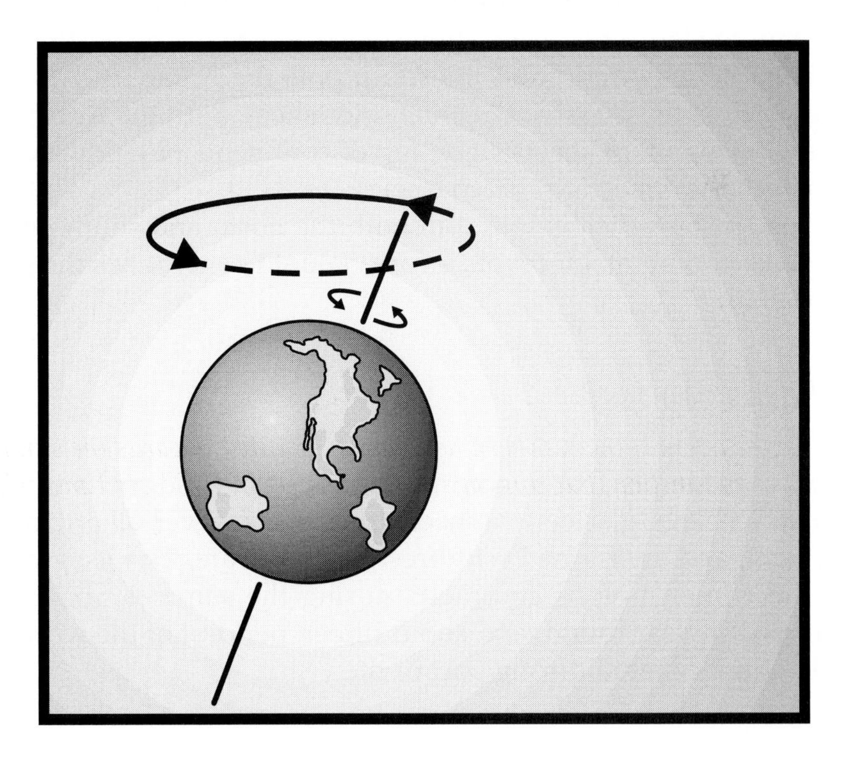

N

NO MORE ICE!

Let's take a look at how global warming is affecting the Earth's coldest regions and ice sheets, collectively called the cryosphere, which is derived from a Greek word meaning frost or cold. It is used to describe the areas of the Earth's surface where water is in a solid form, usually snow or ice.

These areas include sea ice, freshwater ice, glaciers, permafrost and snow.

The Earth's polar icecaps, found at the north and south poles, contain the largest concentrations of ice on Earth. The north pole is home to the Arctic, and the south pole the Antarctic. Also in the north is the massive Greenland ice sheet. Both the Antarctic and Greenland ice sheets sit on top of continents or landmasses, whereas the Arctic is a frozen ocean.

Sea ice, however, is found in both the north and south polar regions, and in total it covers an area about twenty times the size of Canada.

What is sea ice?

Well, it is simply frozen ocean water. It forms and melts in the ocean. Icebergs, glaciers, ice sheets/shelves, however, all originate on land, and are formed with fresh not saltwater. Sea ice grows in the winter months and melts during the summer. Some ice remains all year round, and about fifteen per cent of the world's oceans are covered during part of the year.[1]

Why is ice so important?

Ice has a bright reflective surface, so as sunlight strikes it most of it is reflected back into space. As such, areas covered by ice don't absorb much of the sun's energy, allowing temperatures in the polar regions to remain cool. If higher temperatures melt the ice over time, as is beginning to happen, then more of the sun's energy can be absorbed by the ice-free sea or land, allowing temperatures to rise further.

The term 'albedo' is used to determine how well a surface reflects solar energy. A surface with an albedo of zero means that it is a perfect absorber of the sun's energy, such as a black surface. An albedo of one means that the surface is a perfect reflector, such as a white surface.

Sea ice will reflect about fifty to seventy per cent of the sun's energy. Open sea reflects about six per cent, whereas snow-covered ice about ninety per cent, simply because it's white and therefore has a higher reflective surface.[2]

Much as the Amazon regulates climate by absorbing and storing huge amounts of CO_2, the ice-covered regions of Earth act much in the same way, by regulating temperature and reflecting large amounts of solar energy back into space. If these regions melt, then not only will ocean levels rise but temperatures will also increase.

So how are these regions responding to global warming?

We will look at each of these regions in turn, starting with the Arctic.

The Arctic

The north pole sits right in the middle of the Arctic Ocean, which is fenced in by eight different countries. During the winter the ice extends over the entire ocean and onto the fringes of the land. During the summer, the ice retreats back into the ocean.

Air temperatures in the region have, on average, increased by about 5°C (9°F) over the last 100 years, which is higher than anywhere else on the planet. This has caused Arctic sea ice to decrease by about fourteen per cent since the 1970s.[3]

The local Inuit population have started to notice the warmer summers, the earlier break-up of the ice in spring, and extensive areas of melting permafrost in places like Alaska and Siberia. This in turn is affecting their hunting season, foundations of properties and other infrastructure in the region.

Arctic sea ice has been measured by the National Snow and Ice Data Centre (NSIDC) and NASA, using satellite data, and the findings are that massive reductions in sea ice are occurring at the end of the northern summer.

The sea ice extends to about 15,000,000 square kilometres (5,792,000 square miles) during winter, and down to an average 7,000,000 square kilometres (2,703,000 square miles) during the summer.[4] It therefore loses just over fifty per cent of ice cover after the summer melt season. The annual average extent of Arctic sea ice has decreased by about three per cent per decade since about 1980, which is the equivalent of an area of about 750,000 square kilometres (289,575 square miles). The amount of ice left after the summer melt is also decreasing by about 7.7 per cent each decade.[5]

NSIDC measures Arctic sea-ice extent, or the area of ocean that is covered by at least fifteen per cent ice, which typically reaches its minimum in September, at the end of the summer melt season. On 21st September 2005, the five-day running mean sea-ice extent dropped to 5,320,000 square kilometres (2,054,000 square miles).

The 2006 melting season came to a close on 14th September. On this day, known as the sea-ice minimum, sea ice covered 5,700,000 square kilometres (2,200,000 square miles) of the Arctic, the fourth lowest of the twenty-nine-year satellite record for a single day, but slightly better than 2005's figures.

What about 2007? You guessed it, NSIDC data reveals that Arctic sea ice during the 2007 melt season plummeted to the lowest levels since satellite measurements began in 1979. The September sea-ice minimum went down to 4,130,000 square kilometres (1,594,000 square miles), the lowest September on record, shattering the previous record for the month, set in 2005, by twenty-three per cent.

Computer models have predicted the Arctic will be ice-free in the summer months from 2080 if the overall warming trend continues.

In March 2007, a fire onboard the British nuclear submarine HMS *Tireless* forced it to the surface. Two sailors died in the explosion. The Navy had been conducting tests under the Arctic and the data retrieved indicated that the summer Arctic sea ice may actually be gone by as soon as 2020. This however appears to be a worst-case scenario.

Arctic sea ice is about 2 to 3 metres (6.5 to 9.8 feet) thick on average, so a loss of 7,000,000 square kilometres (2,703,000 square miles) times 2.5 metres (8.2 feet) (thickness) is a considerable amount of water. Melting sea ice, however, does not necessarily add much to sea-level rise when it melts, much like melting ice cubes in a glass do not cause the glass to overflow. Melting glaciers and ice-covered continents however are a different matter and when they melt, sea levels will rise.

Further evidence of Arctic changes comes from the annual Canadian seal cull, which occurs about April each year. Many thousand fewer seals will be killed this year as 2007 has been exceptionally warm, which has caused the sea ice to break up. Unfortunately for the baby seals they need the solid ice to live on for the first couple of weeks after they are born, as they cannot swim. Thousands of seals have died after they have fallen off the melting and cracking ice.

A new NASA-led study found a twenty-three per cent loss in the extent of the Arctic's thick, year-round sea ice cover during the past two winters. The scientists discovered less perennial sea ice in March 2007 than ever before. This drastic reduction is the primary cause of this summer's fastest-ever sea-ice retreat on record and subsequent smallest-ever extent of total Arctic coverage.[6]

Record summer melting has also meant that the usually frozen Northwest Passage waterway, which connects the Atlantic to the Pacific, has become fully navigable, a fact that may raise tensions between Canada, which maintains that the waterway lies in its territorial waters, and other countries in the region.

The melting ice is also causing problems for the polar bear, which can be found only in the Arctic, whose existence is also under threat as a result. Polar bears will be looked at further in Chapter X.

The Antarctic

While the Arctic is an ocean surrounded by land, the Antarctic is a landmass surrounded by ocean!

During the Antarctic winter approximately 18,000,000 square kilometres (6,900,000 square miles) of ocean are covered by ice. During summer this drops to about 3,000,000 square kilometres (1,158,000 square miles).[7] As the Antarctic is surrounded by ocean, moisture is more abundant than in the Arctic, causing Antarctic sea ice to be covered by thicker snow, thereby increasing the albedo effect.

The ice sheet covers about ninety-eight per cent of the Antarctic landmass, and it is the largest single mass of ice on Earth. It is estimated that about ninety per cent of Earth's freshwater is locked up in the ice, which if melted would raise sea levels by an apocalyptic 60 to 70 metres (196 to 229 feet).[8] Thankfully there is no expectation that this will happen at any time soon, as while the coastal glaciers and ice sheets are melting, the centre of Antarctica appears to be currently cooling and receiving greater amounts of snowfall. In 2005, however, NASA scientists detected extensive areas of snow that had melted in West Antarctica in response to warm temperatures. This was the first widespread melting in Antarctica ever recorded and will be looked at in more detail in Chapter U.[9]

While East Antarctica appears to be fairly stable, the western side is far more dynamic. Two ice shelves (ice floating on the sea) that are fed ice from the inner continental ice have recently started to break up – the Larsen ice shelf on the far western side, and the much larger Ross ice shelf. The collapse of the Larsen B ice shelf has been associated with a sustained warming and rapid thinning of the ice shelf.[10]

In February 2001, NASA's Landsat and Modis satellites took pictures (overleaf), showing the Larsen B ice sheet starting to break up. According to NASA, higher temperatures are causing surface melt water to fill crevasses in the ice sheet, thereby increasing pressure on the ice sheet, causing it to crack right through.

NASA reported a massive iceberg (known as B-15) that broke off the Ross ice shelf near Roosevelt Island in Antarctica in mid-March 2000. Among the largest ever observed, the B-15 is approximately 300 kilometres long (186 miles) and 40 kilometres wide (24.8 miles), an area about twice the size of the state of Delaware. The iceberg was formed from glacier ice moving off the Antarctic continent. News reports in July 2008 indicated that the Wilkins Ice Shelf in Antartica is also about to collapse. Very worrying, when considering this is happening during Antarctic winter.

While polar bears live in the Arctic, it's penguins that make the Antarctic their home!

Is Greenland melting?

Recent scientific discoveries of ancient DNA under 1.2 miles (2 kilometres) of ice confirm that Greenland once looked more like its name suggests, a green land! The samples confirmed that forests and fauna existed there when it was free of ice hundreds of thousands of years ago. The large landmass, which belongs to Denmark, is of course now covered in a huge ice sheet, with the bulk of ice in some places extending over 1.2 miles in height (2 kilometres) from the continent beneath it. In fact the continent has been forced below sea level from the sheer weight of the ice sitting on top of it.

The Greenland ice sheets contain enough water to raise global sea levels by 4.5 to 6 metres (15 to 20 feet) should it melt.[11]

The ice sheet alone contains about twelve per cent of the Earth's ice.[12]

Alarmingly, according to the WWF, and NASA, parts of the Greenland ice sheet are melting at up to 42 centimetres (16.53 inches) a year, and the thinning is affecting higher latitudes than expected. Data obtained from NASA satellites show that Greenland continued to lose its mass at a significant rate through April 2006, and this rate is accelerating.[13]

The GRACE satellite mentioned in Chapter L showed that Greenland lost about 220 cubic kilometres of ice in 2005 (52.7 cubic miles). This is almost half the volume of Lake Erie in the USA. To put this loss into perspective, according to the NASA Earth observatory a cubic kilometre (0.23 cubic miles) of ice is equivalent to a trillion (1,000,000,000,000) litres of water or about 264,000,000,000 gallons, about twenty-five per cent more water than the city of Los Angeles uses in one year!* [14]

Put it another way, on the basis that Los Angeles uses about 198,000,000,000 gallons of water a year (seventy-five per cent of 264,000,000,000), then in 2005 alone Greenland lost ice that equated to an amount of water that would supply a city the size of Los Angeles for about 293 years (264,000,000,000 gallons multiplied by 220 cubic kilometres divided by 198,000,000,000).

According to the NASA Earth observatory satellite images, dating back to the 1970s, one particular glacier in Greenland, the Helheim glacier, has remained stationary for decades. In 2001 however it began retreating rapidly, and moved back almost five miles in four years. Now it has started to retreat more quickly, and has also started to get thinner.[15]

The Helheim glacier is a river of ice that pours from the inland Greenland ice sheet and into the sea. Scientists fear that if other glaciers in Greenland start acting in the same way, it could reduce the time it will take the Greenland ice sheet to melt.

Just recently a new island has been discovered off the coast of Greenland, named Warming Island by the American explorer who discovered it. The island was once covered by and joined to the mainland by a glacier, which has melted away, leaving it surrounded by sea. The discovery is further evidence of the warming climate in the region.

A global temperature increase of 2 to 3°C (3.6 to 5.4°F) could trigger the meltdown of the entire Greenland ice sheet.[16] More recent findings, however, report that the ice sheet could withstand a temperature increase of up to 4.5°C (8.1°F).[17]

The polar regions are far more sensitive to temperature increase. If global temperatures rise by 2 to 3°C, polar-region increase could be much more. Researchers have already identified a 2.4°C (4.4°F) increase in southern Greenland during the last twenty years.[18]

This is extremely worrying given that the Stern Review mentions that

> 'Most climate models show that a doubling of pre-industrial levels of greenhouse gases is very likely to commit the Earth to a rise of between 2–5°C (3.6–9°F) in global mean temperatures. This level of greenhouse gases will probably be reached between 2030 and 2060.'

Further,

> 'If the Greenland or West Antarctica ice sheets began to melt irreversibly, the rate of sea level rise could more than double, committing the world to an eventual sea level rise of 5–12 metres (16–39 feet), over several centuries.'[19]

This could spell disaster for coastal areas and the northern hemisphere's climate, as ironically global warming in this instance may cause an abrupt northern hemisphere cooling if sufficient ice melts. This would cause an influx of freshwater, which may affect the ocean conveyor, or ocean thermohaline, which brings warmer water and air to northern Europe, and this will be looked at in the next chapter.

Earth's glaciers and other snow-covered regions will also be looked at in more detail in Chapter Q, when considering the Qori Kalis glacier, which is part of the Quelccaya ice cap in the Andes.

⋆ A '1' followed by nine zeros in the USA and scientific community is referred to as a billion. In the UK the number would be one thousand million. A '1' followed by twelve zeros in the USA is referred to as a trillion. In the UK the same number would be a billion! The European terms are considered the most logical. Billion or bi-llion, meaning two, has twice as many zeros as a million (twelve). A trillion or tri-llion, meaning three, has three times as many zeros as a million (eighteen). The US/scientific basis has been used for these figures.

Key points

- Ice is important as it has a reflective surface, which causes much of the sun's energy to be reflected back into space, keeping the temperatures in the polar regions cool.
- The Arctic has warmed by about 5°C over the last 100 years.
- Warmer air and sea temperatures as a result of global warming are causing ice in the polar regions to melt. This results in increased warming, as there is less ice to reflect the sun's energy back into space, which will cause more and more ice to melt. This in turn will cause further warming.
- Melting sea ice, unlike melting ice from land, does not contribute much to sea-level rise, just as melting ice cubes in a glass do not cause the water to overflow from the glass.
- Ice in the Arctic, Greenland and Antarctic is melting. The Greenland ice sheet alone contains enough water to raise sea levels by about 4.5 to 6 metres should it melt completely.

---

1 National Snow and Ice Data Centre, <www.nsidc.org>.
2 Ibid.
3 WWF, <www.panda.org> ('Climate change in the Arctic').
4 National Snow and Ice Data Centre, <www.nsidc.org>.
5 Ibid.
6 NASA, <www.nasa.gov> ('Sea ice changes leading to record low in 2007').
7 National Snow and Ice Data Centre, <www.nsidc.org>.
8 Wikipedia, <www.wikipedia.com> (on Antarctic ice sheet).
9 NASA, <www.nasa.gov>.
10 Stern Review on The Economics of Climate Change, Part I.
11 NASA, <www.earthobservatory.nasa>.
12 WWF, <www.panda.org> (on glacier facts).

13 Ibid.
14 NASA, <www.earthobservatory.nasa>.
15 Ibid.
16 Stern Review on The Economics of Climate Change, Part I.
17 National Environment Research Council (NERC).
18 WWF, <www.panda.org>.
19 Stern Review on The Economics of Climate Change, Part I.

The following six photographs show the Larsen/ Ross Ice Shelf Collapse. The Larsen Ice Shelf at the northern end of the Atlantic Peninsula experienced a dramatic collapse between 31st January and 7th March 2002.

All images credit NASA visible earth <www.visibleearth.nasa.gov>.

Above:
The Larsen Ice Shelf as seen on 31st January 2002.

Below:
The Larsen Ice Shelf as seen on 17th February 2002.

Above:
The Larsen ice shelf as seen on 23rd February 2002.

Below:
The Larsen Ice Shelf as seen on 7th March 2002.

Above: Larsen break-up close-up:
Larsen Ice Shelf collapse

Below: Ross Ice Shelf:
Sea ice breaking away in Ross Sea, Antarctica on 9th November 2001.

O

OCEAN THERMOHALINE CIRCULATION

The oceans, which cover about seventy per cent of the Earth's surface, act as major 'climate controllers' due to their capacity to retain large amounts of heat. It actually takes four times the amount of energy to raise the temperature of water by one degree than it does the soil, for example.

The oceans over a long period of time can store and transport heat from one location to another.[1] They also react slowly to the surrounding atmosphere, therefore any changes occurring in the oceans now are likely to have been a result of past climatic influences rather than anything that is going on at present. Likewise anything going on now in the atmosphere will affect the oceans in decades to come.

The oceans also act as large sinks for CO_2. As the oceans warm, however, which is happening now, their capacity to absorb CO_2 diminishes, meaning more CO_2 will end up in the atmosphere, thereby increasing greenhouse gas levels, which in turn will cause further warming (as discussed in Chapter G).

The oceans also contain huge amounts of photosynthesising phytoplankton, and it is estimated that these tiny-celled plants produce up to an incredible seventy-five per cent of Earth's oxygen. The oceans are the Earth's lifeblood, and the currents that transport nutrient-rich seawater around the planet are vital for Earth's health.

What is the ocean thermohaline circulation?

The ocean conveyor belt, meridional overturning circulation (MOC), and ocean thermohaline circulation (THC), are all names used for a large system of deep currents that bring warmer water from the equator to the northern hemisphere, and returns colder water south back to the equator.

The name 'thermohaline' is derived from the process that relies on both temperature (therm) and haline (salt) to drive it.

The Gulf Stream, which is mainly a wind-driven surface current, carries warm waters north along the eastern seaboard of the USA, whereas the North Atlantic Drift driven by the thermohaline conveyor carries warm water northeast across the Atlantic. It then flows south from the European side back down to the equator, where it joins another current, and eventually reaches the Pacific Ocean. Don't expect to hitch a ride on the current, however, as the process takes about 1,000 years to complete![2]

As the warmer water moves to higher latitudes, carried towards Europe by the North Atlantic Drift, colder dry air causes a lowering of the warmer water temperature and evaporates the surface ocean water, leaving behind denser salty water. The denser water has a tendency to sink in a process called deep convection. The warmer water and colder water then mix in certain areas around the Labrador Sea, which is off the coast of Newfoundland, the Weddell Sea in Antarctica, and the Mediterranean Sea in Europe.[3]

The deep-water masses then spread, leading to a formation called North Atlantic Deep Water (NADW).

What's so important about the THC?

The presence of the ocean thermohaline circulation ensures Europe is maintained at a few degrees warmer than it otherwise would be. You only have to look at a map of the world to see that London, England, sits at 51.30 degrees north of the equator, which is a similar latitude to Saskatoon in Saskatchewan, Canada. Saskatoon is usually covered in a blanket of snow for about five months of the year during winter. Glasgow in Scotland is on the same latitude as the Hudson Bay in Canada. While it may get quite chilly in Glasgow, the latter is covered in sea ice for much of the year. When you consider this, the importance of the North Atlantic Drift cannot be underestimated in helping to maintain Europe's moderate climate. The ocean circulation is also very important, as mentioned above, in maintaining the transport of nutrients around the world's oceans and for the sake of all marine life.

The premise for the 2004 movie *The Day After Tomorrow* was a collapse of the THC, but in the movie this happened over a period of weeks, which of course is very unlikely, we hope!

Is the ocean thermohaline being affected by global warming?

According the Stern Review scientists seem to think that 'maybe'

> 'Climate change could weaken the Atlantic Thermohaline Circulation, partially offsetting warming in both Europe and eastern North America, or in extreme case causing significant cooling. The warming effect of greenhouse gases has the potential to trigger abrupt, large scale and irreversible changes in the climate system. One example is a possible collapse of the North Atlantic Thermohaline Circulation (THC). In the North Atlantic, the Gulf Stream and North Atlantic drift (important currents of the THC) have a significant warming effect on the climates of Europe and parts of America. The THC may be weakened, as the upper ocean warms and/or if more fresh water (from melting glaciers and increased rainfall) is laid over the salty seawater. No complex climate models predict a complete collapse. Instead these models point towards a weakening of up to half by the end of the century. Any sustained weakening of the THC is likely to have a cooling effect on the climates of Europe and Eastern North America, but this would only offset a portion of the regional warming due to greenhouse gases. A recent study using direct ocean measurements (the first of its kind) suggests that part of the THC may

> have already weakened by up to 30 per cent in the past few decades, but the significance of this is not yet known.'[4]

The fact the both the Arctic and Greenland ice sheets appear to be melting may be contributing to the slowing down of the THC. As more freshwater from the melting ice enters the oceans, this helps disrupt the thermohaline process as freshwater mixes with more saline water, thereby affecting its density. If the dense, salty water is diluted by the fresh meltwater from ice it will have less propensity to sink, which could be an explanation as to why the THC is slowing down, if the scientific findings are correct.

Ongoing scientific research

A team of scientists based at the National Oceanography Centre at Southampton University have been researching the flow of current in the Atlantic Ocean at latitude twenty-five degrees north. They have discovered that it has slowed down by thirty per cent, by comparing surveys taken in 1957, 1981, 1992, 1998, with a recent one taken in 2004.

Measurements have been taken by ships at regular intervals across the Atlantic to measure ocean flow. However, due to the importance of the data and the consequences for European climate, if the thermohaline does indeed slow or grind to a halt, continuous observing systems have now been put in place to monitor the circulatory overturning.[5]

The Natural Environmental Research Council (NERC) has funded the programme called RAPID, which the scientists at Southampton have been co-ordinating. The programme was set to run until 2008, giving four years' worth of data to analyse. The programme has recently been extended to 2014, however. The results from the first year's data indicate that the previous publicised observations may lie within natural variability.[6]

The data collected from the monitoring arrays that have been set up may give scientists a better idea as to whether the ocean thermohaline is indeed being affected by global warming, and whether the initial findings that show it to be slowing are indeed correct. Only time will tell!

If so, then ironically global warming may cause cooling in North America and western Europe, possibly causing a significant drop in temperatures. While scientists, thankfully, do not expect the thermohaline to shut down completely, it appears from historical proxy data that this has happened before in the Younger Dryas period, which caused a relatively sudden drop in temperatures as discussed in Chapter H.

In the next chapter, 'Population Growth', we will look at another factor relevant to global warming, as of course Earth's population is ever growing, and as a result will have both an increasing affect and be increasingly affected by global warming in the future.

Key points

- The ocean thermohaline circulation (OTC) is the name given to a system of currents that bring warm water from the equator to the northern hemisphere while taking cold water away.
- The Gulf Stream is mainly a wind-driven surface current, whereas the North Atlantic Drift is driven by the thermohaline circulation.
- The current is very important to Europe's climate, as without it temperatures would be a few degrees centigrade lower.
- Scientists think that the current could be affected by climate change and global warming, as it appears to have slowed by thirty per cent, though this change could be within natural variability ranges.

---

1 NASA Goddard Institute for Space Studies, <www.giss.nasa>.
2 Ibid.
3 Ibid.
4 Stern Review on The Economics of Climate Change.
5 Mongabay, <www.mongabay.com> ('Could the Atlantic current switch off?').
6 NERC, <www.nerc.ac.uk> (NERC open meeting, 31st January 2007).

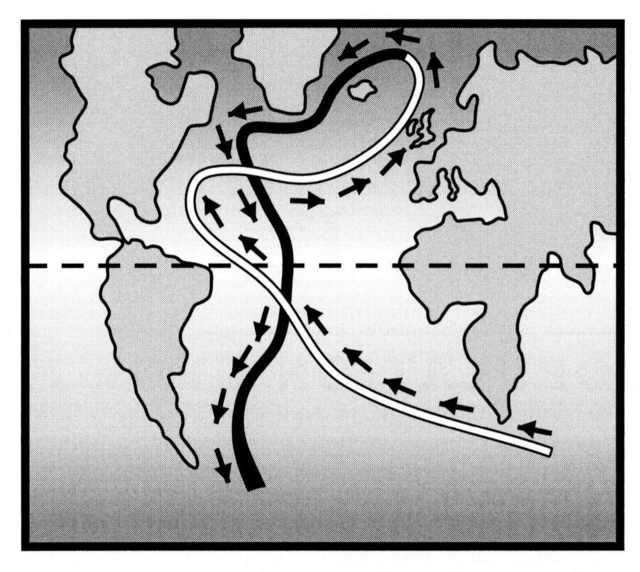

Above: Warm equatorial water is brought to northern latitudes by the Ocean Thermohaline Circulation, on currents called the Gulf Stream and North Atlantic Drift.

Below: The system is driven by relatively colder dry air in northern latitudes evaporating the warmer surface water, leaving cooler, denser water, which then sinks, a process that drives the conveyor-like circulation.

P

POPULATION GROWTH

Population and population growth have a profound affect on Earth's resources, environment, and ultimately climate. As the Earth's population increases, there will be a need for more energy, more electricity generating power plants, cars, and deforestation to make room for living space.

World population now stands at about 6,500,000,000 people. It took all of human history until 1830 for world population to reach 1,000,000,000. One hundred years later world population stood at 2,000,000,000. By 1960 it had hit 3,000,000,000. By 1975 population grew to 4,000,000,000, then only twelve years later in 1987 it increased to about 5,000,000,000. Today world population is, as I said, about 6,500,000,000, and growing at about 80,000,000 a year. At current rates world population will total about 9,000,000,000 by 2050.[1]

The major stresses on populations caused by global warming and climate change will be lack of freshwater, higher temperatures and adequate food production.

Water

Water is essential for the existence of all life forms, and over half of the Earth's drinking water is taken directly from rivers or reservoirs, while the rest comes from groundwater. Almost seventy per cent of water withdrawal is for irrigating crops and supplying food. A further twenty-two per cent is for industrial purposes, including manufacturing and cooling of thermoelectric power generation. Just eight per cent is used for domestic household use and businesses for drinking, sanitation and recreation.[2]

The availability of water in different regions of the world will become a more pronounced issue. Areas that are already relatively dry, like the Mediterranean basin, parts of southern Africa and South America are likely to experience further decreases in water availability. However, regions such as southern Asia, northern Europe and Russia are likely to experience increases in water availability, perhaps up to ten to twenty per cent if there is an increase in temperatures of about 2°C (3.6°F), according to several climate models.[3]

The reason is more rain at high latitudes, less rain in dry subtropical regions, and hotter land surface temperatures, which will induce more powerful evaporation and hence more intense rainfall, but with an increased risk of flash flooding.

As the water cycle intensifies, billions of people will lose or gain water depending on their geographical location. The effects of rising temperatures, against a background of increasing population, are likely to cause changes in the water status of billions of people. According to one study, a temperature rise of 2°C (3.6°F) will result in 1–4,000,000,000 people experiencing growing water shortages, predominantly in Africa, the Middle East, southern Europe and parts of South and Central America.[4]

At the same time 1–5,000,000,000 people, mainly in southern and eastern Asia, may receive more water, but this will arrive during the wet season and will be useful only for alleviating water shortages in the dry season, if storage of water can be created.[5]

Populations that depend heavily on glacial meltwater will have difficulty maintaining supplies in the dry season. The Indian subcontinent, 250,000,000 people in China, and tens of millions in the Andes will be severely affected as global warming causes the glaciers to melt. This will be looked at in more detail in Chapter Q, which deals with glaciers.[6]

Temperature

The Earth has already warmed by 0.74°C (1.33°F) since 1900, and is committed to further warming through past emissions over the coming decades. On current trends, average global temperatures could rise by 2–3°C (3.6–5.4°F) within the next fifty years or so, with even higher temperature increases by the end of the century, if emissions grow.[7]

In northern latitudes (Europe, Russia, Canada, the USA), global warming may mean fewer deaths overall with populations spared cold-related deaths in the winter, compared to heat-related deaths in the summer.

In cities, however, heatwaves will become increasingly dangerous, as regional warming together with the heat island effect (where cities concentrate and retain heat) leads to extreme temperatures and more toxic air pollution incidents.[8]

This is a worrying fact as according to the Population Institute, as of 2007 more than fifty per cent of the world's population will live in urban areas for the first time in human history.

Food

Declining crop yields are likely to leave hundreds of millions of people without the ability to produce or purchase sufficient food, particularly in the poorest parts of the world.

Food production will be particularly sensitive to climate change, because crop yields depend in the main on prevailing climate conditions (temperature and rainfall patterns). Low-level warming in mid to high latitudes (USA, Europe, Australia, Siberia and some parts of China) may improve the conditions for crop growth by extending the growing season. Further warming, however, will have increasingly negative impacts as damaging temperature thresholds are reached more often, and water shortages limit growth in regions such as southern Europe and the western USA.[9]

If temperatures increased by 4°C (7.2°F) entire regions may be too hot and dry to grow crops. Agricultural collapse across large areas of the world is possible if temperatures were to increase by 5 or 6°C (9 to 10.8°F) above present levels, but evidence for this occurring is still limited.[10]

The Earth's increasing population will therefore put enormous strain on an already changing environment, and of course cause further global warming, as increased resources are needed to sustain development and the population in general. Perhaps the greatest risk to peace and stability comes from the pressure that global warming will put on populations, as our water, temperature and food all become affected by the Earth's changing climate, forcing populations to migrate, and into possible conflict over Earth's resources, food, water and land.

In the next chapter we will look further at the Earth's tropical glaciers, which will be affected as they start to melt and influence the freshwater supplies of millions of people.

Key points

- Earth's population is now at about 6,500,000,000 people. Only twenty years ago it stood at 4,000,000,000. By 2050 it could be 9,000,000,000!
- Population growth will contribute to global warming as demand for energy increases, which in turn will cause more fossil-fuel pollution. At the same time, global warming will put enormous pressure on Earth's population as water, temperature and food all become affected.

---

1 The Population Institute, <www.population.org>.
2 Stern Review on The Economics of Climate Change, Part II.
3 Ibid.
4 Ibid.
5 Ibid.
6 Ibid.
7 Ibid.
8 Ibid.
9 Ibid.
10 Ibid.

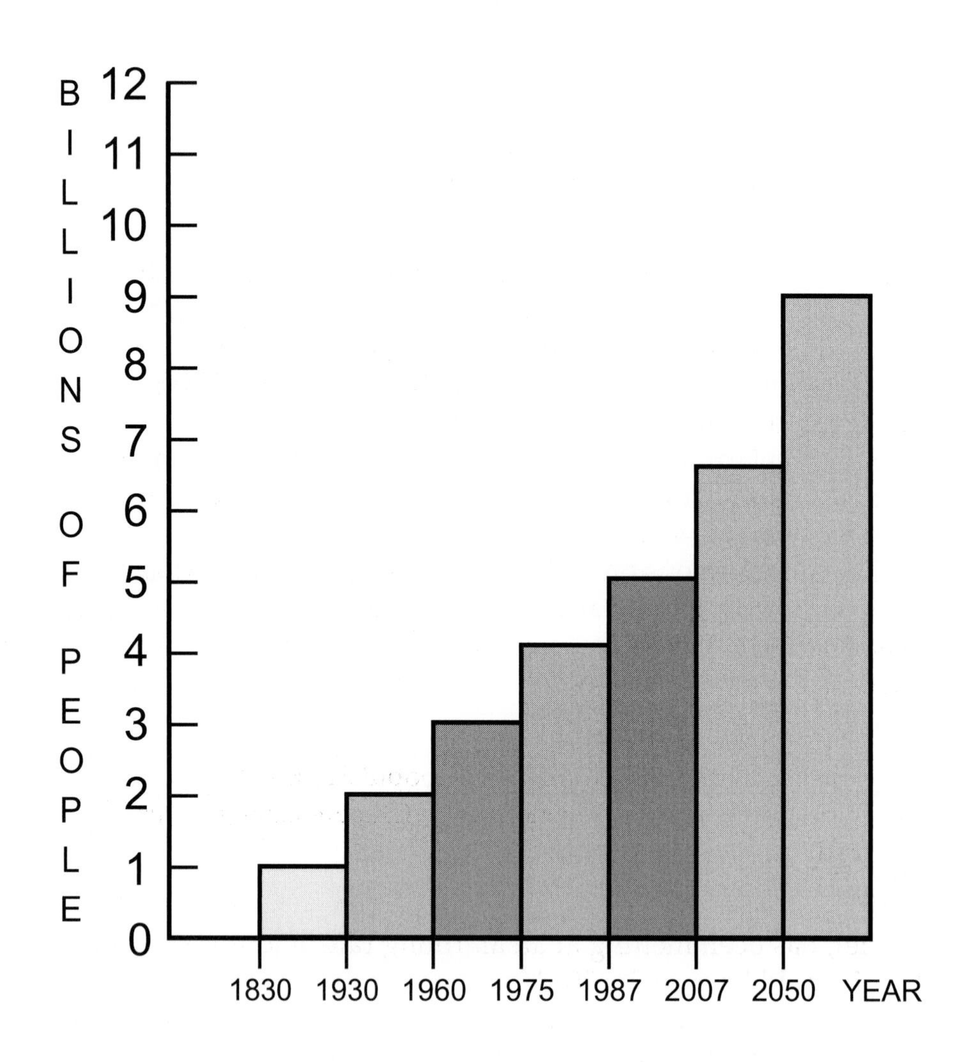

Population graph showing population growth 1830–2050.
Credit Population Institute figures.

Q

QUELCCAYA ICE CAP

While the vast majority of Earth's ice is locked up in Antarctica and Greenland, there are countless numbers of glaciers in the tropical regions of Earth, which supply almost a third of the Earth's population with fresh drinking water. That about seventy per cent of Earth's population lives in the tropics, a region which contains about fifty per cent of Earth's landmass, means serious consequences for people in these regions, as glaciers start to melt.

Glaciers act as the Earth's freshwater reservoirs. Collectively they cover an area the size of South America. While glaciers have been retreating worldwide since the end of the Little Ice Age (about 1850), they have in recent decades been melting at rates that far exceed historical trends.

A good example is the Quelccaya Ice Cap, found in the Andes mountain range in Peru, which is the largest glaciated region in the tropics. The ice cap is found at an altitude of about 5,500 metres (18,044 feet).

One of the largest glaciers making up the ice cap, the Qori Kalis glacier, has been melting at an alarming rate, and it is estimated that it could be gone by 2012.[1]

Glaciologists and researches from the Ohio State University, led by Dr Lonnie Thompson, have been studying the area since 1974, and have found that tropical glaciers around the world are in retreat. Dr Thomson has found that Peru's Qori Kalis glacier is melting at a rate of some 60 metres (200 feet) per year.[2]

The Himalayas, and the snows of Mount Kilimanjaro in Tanzania, are also melting.

'No' snows of Kilimanjaro!

According to Dr Thompson,

> 'Kilimanjaro is behaving just like Mount Kenya and the Rwenzori, both also in Africa, as well as glaciers in the Andes and the Himalayas. This widespread retreat of mountain glaciers may be our clearest evidence of global warming as they integrate many climate variables. Most importantly they have no political agenda.'[3]

You have only to look at the photographs of Kilimanjaro in this chapter to see the difference in snow cover on the mountain's summit over a seven-year period.

The Peruvian government estimates that the country's glaciers have shrunk by more than twenty per cent.[4]

According to the Stern Review, tropical Andes glaciers have reduced in size by nearly twenty-five per cent during the last thirty years. Tens of millions of people in the Andes will be affected by this. In addition, about 250,000,000 people in China also rely on glacial meltwater, especially in the Himalayas, to maintain water supplies during the dry season. These supplies will dry up in the long run if the glaciers melt completely.[5]

Simulations have projected that a 4°C (6.4°F) rise in temperatures would cause almost all of Earth's glaciers to melt.[6]

Goodbye glaciers?

1. Northern Andes region. This region contains the largest concentration of tropical glaciers, which is receding rapidly, with the melting accelerating since the 1990s.
2. Himalayan glaciers have retreated and thinned over the last thirty years, with accelerated losses over the last decade. A recent newspaper article confirmed that the original Everest base camp is now considerably lower than it was originally, proof that Everest's glaciers are melting and receding.

3. Nearly all glaciers surveyed in Alaska are melting. Fifty per cent of all water flowing into the oceans, globally, is due to melting glaciers in Alaska.
4. Glacier melting in the European Alps has accelerated since 1980. European alpine glaciers have lost fifty per cent of their volume since 1850. By the end of the twenty-first century they will have lost seventy-five per cent.[7]

It seems the loss of Europe's ski regions, while tragic for ski enthusiasts and the economy alike, will be nothing compared to the problems that will be encountered by millions of people as melting glaciers affect water supplies in the regions mentioned above.

In the next chapter we will look at renewable energy and the various ways power can be generated without having to rely on greenhouse-gas-polluting fossil fuels. Renewable energy sources will play an important role in the future of energy production if greenhouse gases are to be reduced to avoid runaway global warming.

Key points

- ➢ Glaciers located in Earth's tropical regions supply about thirty per cent of Earth's population with fresh drinking water.
- ➢ While glaciers have been in general retreat since the end of the Little Ice Age, circa 1850, their loss has been accelerating in recent decades.
- ➢ Glaciers in the Andes have reduced in size by nearly twenty-five per cent in the last thirty years, which will affect the water supply of millions of people.

1 Mongabay, <www.mongabay.com>.
2 Ibid.
3 Ibid.
4 Ibid.
5 Stern Review on The Economics of Climate Change, Part II.
6 WWF, <www.panda.org> (on glaciers at risk).
7 Ibid.

Mt Kilimanjaro, 1993. Some scientists believe the snowcap of Mount Kilimanjaro will be gone in two decades. Researchers say the ice fields on Africa's highest mountain shrank by eighty per cent in the past century. The snowcap formed some 11,000 years ago. The Landsat satellite captured these images of Kilimanjaro, 17th February 1993 (top) and 21st February 2000 (bottom).

Credit NASA visible earth, <www.visibleearth.nasa.gov>.

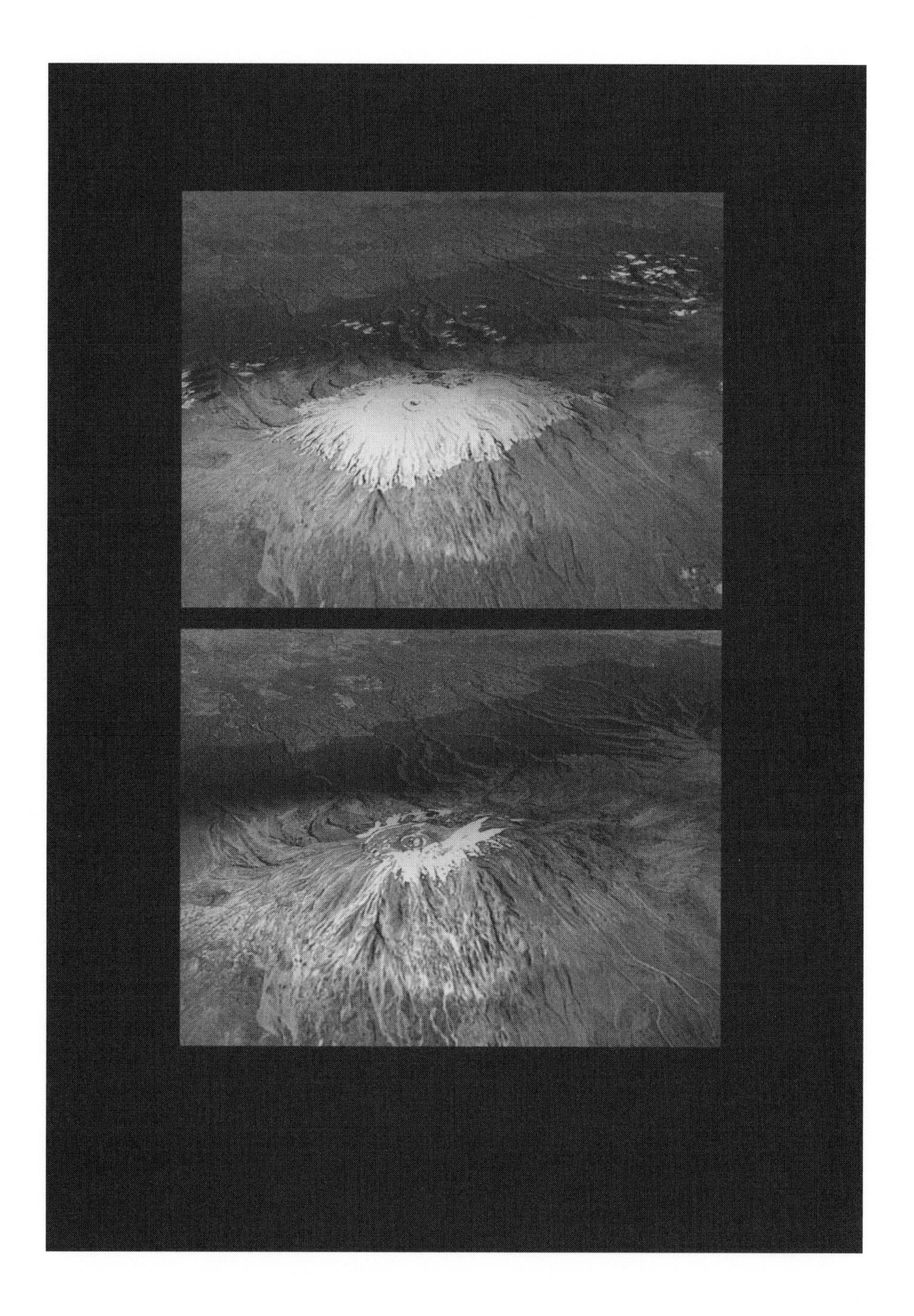

R

RENEWABLE ENERGY

Renewable energy is any energy derived from resources that can be regenerated. Natural sources, such as solar, wind and water, fall into this class, as opposed to energy obtained from burning fossil fuels, such as coal, oil and gas.

The essential difference is that renewable energy is virtually non-polluting if applied correctly, which means energy production from renewable sources does not add global-warming greenhouse gases to the atmosphere, unlike the bulk of fossil-fuelled coal power stations operating at the moment.

The main sources of renewable energy are

1. Solar power
2. Wind power
3. Hydropower
4. Geothermal
5. Biomass

We will take a look at each energy source in turn and explain how it is obtained and applied to produce electricity for power.

In order to help understand and compare the power each energy source produces, we must first understand how energy is measured. Energy is measured in watts per second, and one watt is equal to the power that in one second gives rise to one joule of energy.[1]

Watt's all this about?

The rate at which electricity is used is measured in watts, so a 100-watt light bulb uses 100 watts. A 1,500-watt vacuum cleaner uses 1,500 watts and so on.

To understand how much energy is being used, the length of time an appliance or light is on must be considered. A 100-watt bulb would use 100 watt hours if it was left on for one hour, or 0.1 kilowatt hours (100 divided by 1000).

A 100-watt light bulb left on for ten hours would use 1,000-watt hours, or one kilowatt, (100 x 10 = 1,000 watts or one kilowatt).

A typical household may consume an average of about 1,000 watts. This would equate to about 20,000 to 24,000 watt hours per day, or 20 to 24 kilowatt hours each day. Obviously energy use varies considerably, but if you left one 100-watt light bulb on all day and all night, you would use 24 x 100 = 2,400 watt hours (2.4 kilowatt hours) during that time period. That's not taking into account all the other appliances and lights that may be on or running at the same time!

So, considering an electricity tariff of say 9.8 pence per kilowatt hour (approximate UK off-peak tariff) then to leave a 100-watt light bulb on for twenty-four hours would cost about 23.5 pence (100 x 24 = 2,400 watts divided by 1,000 = 2.4 kilowatt hours x 9.8 = 23.5 pence).

One megawatt is equal to 1,000,000 watts, enough to supply about 1,000 houses with electricity.

A medium-sized coal-fired power station generates about one gigawatt of power or 1,000,000,000 watts.

Two of the UK's largest power stations, Longannet and Cottom, generate about two gigawatts of power each and produce about 8,417,779 and 8,068,565 tonnes of CO_2 annually respectively![2]

Did you know that 1 tonne of coal when burned produces about 3.7 tonnes of CO_2?[3]

Okay, that's the complicated stuff explained! We will start by looking at energy that comes from the star at the centre of our solar system, our sun.

Solar energy

Solar energy is energy derived from the sun. The potential of solar energy is immense. The sun's energy can be harnessed in many ways. Photovoltaic cells or solar panels can be used to convert solar radiation into electricity. Solar thermal collectors can convert solar radiation to heat water, and solar thermal power plants use the sun's energy to heat up water, or oil-based liquids, which can then convey steam in order to power a turbine or sterling engine.

Theoretically the Earth's entire present energy consumption could be met by an array of photovoltaic solar cells covering an area of 700 square kilometres (270 square miles)![4]

It would of course be impractical to do this, but it just shows the power of the sun, if energy could be harnessed in this way.

In the USA, or especially California, Governor Arnold Schwarzenegger recently announced the California Solar Initiative, which sets a goal for California to produce 3,000 megawatts of solar-powered electricity by 2017, equivalent to three medium-sized coal-fired electric power stations.[5]

The USA also has generous tax incentives, up to $2.50 per watt on solar photovoltaic roof panels, which means that up to fifty per cent of the cost of a solar panel system can be financed – much better than the grant system operating in the UK at present.

Many companies are taking advantage of the rising demand in photovoltaic roof panels. Nanosolar Inc, which is a California-based company, is building one of the largest solar cell producing facilities. The annual output of the facility will have the energy-generating equivalent of one coal-fired power station.[6]

Another California-based company, Solar Integrated Technologies, is producing photovoltaic panels for commercial use and has recently taken orders for schools, Tesco's stores and other large flat-roofed buildings in California and Europe.[7]

Solar panels used for heating up domestic hot water supply can be purchased fairly cheaply, about £1,500 in the UK from DIY store B&Q. Fitting costs would be in addition to the purchase cost, however. A hot-water tank is required to store the water once it has been heated by the panels, which is a problem as most properties in the UK have combination boilers. Therefore, additional costs are mostly required in fitting a storage tank!

Thermal power plants are going to have to be a major provider of electricity in the future if governments seriously intend cutting down on greenhouse gas emissions. Spain has just completed a plant near Seville. It consists of about 600 massive mirrors, which focus the sun's energy on the top of a forty-story tower, where it heats up a huge water tank/boiler. This produces huge amounts of steam, which is used to drive a turbine. The plant produces about eleven megawatts of power, enough to supply perhaps 11,000 homes with electricity. Expansion is already taking place, which will eventually provide enough power for the entire population of Seville, all with zero greenhouse gas emissions!

In the future, huge areas with similar solar thermal plants could be set up in the Sahara desert, for example, to supply Europe and other areas with pollution-free electricity. This possibility is apparently being looked into.

Wind power

The wind energy industry is growing rapidly. Since 1998 it has grown sevenfold, according to the World Wind Energy Association (WWEA). As of December 2006, about 74,000 megawatts of electricity are being generated worldwide from wind turbines.[8]

In Europe alone the WWEA anticipates production rising to 160,000 megawatts by 2010, an almost 120 per cent increase, which will produce the equivalent electricity of 160 medium coal-fired power stations. Germany and Spain lead the way, with France and Portugal following closely behind.

The USA also added 2,424 megawatts of new capacity in 2005, with Canada doubling its capacity to 683 megawatts of power.[9]

The world's largest wind-power developers, Acciona, will be developing its first wind turbine plant in the state of Iowa, USA, increasing the USA's wind generating electricity capacity further. Acciona has two other plants in Spain and one in China.[10]

Of all the renewable energy technologies, the cost of wind energy comes closest to that of fossil-fuel production, and continues to fall, meaning huge worldwide expansion potential.

The Asia-Pacific area, however, remains the most dynamic region, with wind energy producing a total capacity of 7,000 megawatts, equivalent to seven medium coal-fired power stations. Both India and China are leading forces, with India now ranked fourth in the world, and China following close behind in sixth place.[11]

Global wind energy potential is massive. Theoretically onshore potential alone, in sites with a wind speed of at least five metres per second at ten metres high, would produce the equivalent electricity of thirty-five times the total amount of current world electricity consumption![12]

A single 1.5 megawatt wind turbine produces enough electricity for about 1,000 houses, and throughout its service life can keep 80,000 tonnes of brown coal out of the atmosphere. This is a coal pile covering an entire football pitch and rising almost as high as the wind turbine itself![13]

Hydropower

Electricity derived from the water includes hydroelectric power, tidal and wave power.

Hydroelectric power

On a worldwide basis, hydroelectric power contributes about nineteen per cent of electricity generation. The power is usually harnessed by the construction of a dam, which holds back water, which can then be released in a controlled fashion to drive a turbine, which then creates electricity.[14]

Tidal power

Electricity can be generated using the power of the tides, especially in narrow river estuaries or sea straits. The oldest such power plant was set up in 1965 in the River Rance near St Malo in France, which has a generating capacity of 2,400 megawatts. It has worked without failing ever since![15]

There have been decades of continuing debate and proposals about such a scheme in the Severn Estuary in the UK, which has one of the largest tidal waters in the world. However no such scheme has yet got off the ground. Discussions about such a project appear to have surfaced again in the media very recently.

Wave power

Finally kinetic energy can be captured from the sea by harnessing the power of waves.

In February 2007, the world's largest wave farm, which will be based off the Scottish coast in the UK, was given the go-ahead, generating three megawatts of power.

Geothermal energy

Energy from geothermal sources is simply energy from the heat of the Earth. The centre of the Earth is as hot as the surface of the sun, somewhere in the region of 5,500°C (9,932°F).[16]

At 3-metres depth (9.84 feet) the temperature remains at a constant 10.16°C (50.28°F) throughout the year. On average the temperature increases by about 3°C (5.4°F) for every 100 metres (328 feet) of depth.[17]

Buildings can be both heated and cooled by tapping in to the top 15 metres (49.2 feet) or so of the Earth's surface.

How does this work?

During the winter, when the ground is warmer than the buildings above, liquid stored in a network of pipes beneath the ground absorbs heat, which can be pumped to the surface and used to warm the buildings through a series of heat exchangers and collectors.

During the summer, when the buildings above are warmer then the ground below, a pump can transfer heat from the buildings back into the ground, or cools via an air conditioning system.[18]

Deep geothermal energy can also be obtained from great depths via deep boreholes. Thermal groundwater can be used directly in hydrothermal power plants to generate both electricity and heat.

These types of plants produce 1,000 to 2,000 times less CO_2 than fossil-fuel plants, and the electricity produced is available ninety per cent of the time compared to a fossil-fuelled plant, which produces electricity sixty-five to seventy per cent of the time.[19]

These types of plants, however, are still in their infancy. In certain countries, which have favourable geological conditions, the technology is already established. Iceland, the USA, Indonesia, New Zealand, Mexico and Italy all have geothermal programmes up and running.

The Nesjavellir geothermal plant in Reykjavik supplies over fifty per cent of the population with energy from geothermal water processed from the plant, producing only steam instead of nasty polluting greenhouse gases in the process.

Biomass

Biomass is discussed in detail in Chapter B, so won't be discussed further here.

What are the downsides?

While having zero-emission energy is fantastic for preventing further pollution and global warming, there are of course some practical difficulties with renewable energy generation.

Lack of sun

Photovoltaic solar panels are more efficient in countries that get more sun/sunlight hours. Notwithstanding this, northern Europe has embraced the concept with large photovoltaic plants operating in both Germany and Portugal. Australia is currently building the largest photovoltaic plant in the world, which when complete will supply zero-emission electricity to thousands of households.

There are literally hundreds, if not thousands of companies springing up to provide photovoltaic panels for commercial electricity generation, and more and more businesses are installing this technology to generate power.

Expense

As for the private individual, however, the cost of installing enough photovoltaic panels in your house (twenty or so one-metre-square panels for household electricity production) will have to come down considerably before it becomes worthwhile to generate enough electricity for day-to-day household demand. Governments will have to start providing proper subsidies, and perhaps follow the California model in the USA, making solar electricity generation affordable to the general public.

Photovoltaic cells are used to generate electricity and should not be confused with solar panels or thermal collectors, which use the sun's energy to heat water.

Using solar panels to heat up the domestic water supply is more affordable, perhaps in the region of £4,000 ($8,000) installed, and prices will continue to fall. This can of course enhance the value of a property in any event. So, even if it takes a while to recover the cost of the solar panels by having reduced or zero water-heating bills, the panels themselves will constitute a desirable asset for any property, when it is time to sell. The slight problem is however that most properties, certainly in the UK, are fitted with combination boilers, which means there is no water tank. Solar heated water systems require a tank to operate, therefore this will be required in addition to the combination boiler in any event! Under new regulations, however, all new housing now being built will require condensing boilers to be fitted, usually in the attic, as they give off steam, which also have a hot-water tank, and so would be more compatible with solar-heated systems.

Enough wind?

As for wind turbines, there are various arguments that have been made, including the fact that they are not aesthetically pleasing, make too much noise and even that they are a danger to birds!

These gripes are surely ridiculous when considering the thousands of tonnes of CO_2 and other chemicals each turbine keeps from the atmosphere during its operating lifetime, not to mention the smell and impact a big power station has on the countryside.

A reasonable argument in relation to the efficiency of wind farms exists, however, on the basis that the location of the wind turbines affects the capacity at which they work. For the turbines to be efficient and meet UK government targets – ten and fifteen per cent of electricity generated from renewable energy sources by 2010 and 2015 respectively – each turbine needs to work at thirty per cent capacity.

The energy regulator OFGEM has looked at the data and it appears that the only regions in the UK where wind farms operate at thirty per cent capacity or greater are in Scotland.

The most appropriate locations for wind farms are in the windiest locations, and these tend to be offshore. However, this does not mean that located properly they can't contribute significantly to UK and world energy needs. Indeed, it has recently been announced that the world's largest offshore wind farm has got the green light, and will be erected off the south coast of Britain in a development called the Atlantic Array. When operational this could prevent a few million tonnes of CO_2 from being added to the atmosphere, by producing electricity in place of a coal-fired power station that otherwise would have been required. Recently the Ministry of Defence has expressed concerns that the wind turbine's profile interferes with its radar, creating a security threat to the UK, but these concerns should hopefully be overcome!

Hydroelectric dams have been criticised for having a large impact on the natural environment, and sometimes for exporting much of the electricity produced to benefit areas away from the dam itself. This is of course not very fair to the local population, who have to put up with the dam, but perhaps not benefit from the electricity it produces!

What about nuclear power?

Electricity is generated inside a nuclear reactor as nuclear fission takes place. This is done by a controlled chain reaction. Fissile material – such as uranium or plutonium – is bombarded by neutrons, splitting the uranium or plutonium atoms. The atoms split into two or three smaller nuclei in a chain reaction, which creates energy. The energy is then used to heat water, which produces steam, which then powers a turbine, much the same as in a coal-fired plant. Nuclear power stations do not generate greenhouse gases. However, nuclear energy is not legally classified as renewable energy, as it would allow it to be applicable for development aid in various jurisdictions. At present about sixteen per cent of the world's electricity is generated from the 422 nuclear power stations, which operate in 30 different countries according to the International Atomic Energy Agency (IAEA).[20]

While there had been a gradual phase-out of new nuclear power generating capacity during the last few decades, it does seem, with the necessary phase-out of fossil-fuel energy production, that nuclear energy production will have to grow to fill the gap, together with true renewable resources. There are of course major objections by various organisations, including the WWF and Greenpeace, to any expansion of nuclear power electricity generation. However, it may be the only answer if countries like India and China are to prevent runaway greenhouse-gas emissions from their fossil-fuel energy plants, as their economies and populations continue to expand at current rates.

All in all, however, energy from renewable sources is ready to take off in a big way, as climate change from global warming becomes more evident, and causes greater impact on the environment. As technology in this field improves and becomes more efficient, using the Earth's natural, pollution-free energy sources will become a necessity in reducing reliance on fossil-fuel energy production. Indeed, according to the IPCC, it is essential that more of the world's power is derived from renewable sources, in order to avert catastrophic global warming caused by increased CO_2/greenhouse gas levels.

We will now look at the ultimate power plant that drives the Earth's entire weather system and powers all renewable energy sources, the sun, and consider its role in global warming and climate change.

Key points

- Renewable energy is energy derived from resources that can be regenerated, such as from natural sources.
- A typical home consumes on average about 1,000 watt-hours, that is 1,000 watts each hour.
- A medium-sized coal-fired power station produces about one gigawatt of power, enough to supply about 1,000,000 homes.
- To produce this much power, a coal-fired power station would emit about 4,000,000 tonnes of CO_2 annually.
- Wind power is the fastest growing renewable energy source and the world's largest wind farm is being built off the south coast of England.
- An array of photovoltaic solar panels covering an area of 700 square kilometres could theoretically supply Earth's entire current energy needs.

---

1 'Watt', in *A Dictionary of Weights, Measures, and Units*, Oxford University Press, 2002, 2004. Answers.com, 2nd December 2007, <http://www.answers.com/topic/watt>.
2 WWF, 'Dirty Thirty' May 2007 report.
3 <www.treehugger.com>.
4 WWF, <www.panda.org>, on solar power.
5 Californian Solar Initiative.
6 Nanosolar, <www.nanosolar.com>.
7 Solar Integrated Technologies, <www.solarintegrated.com>.
8 World Wind Energy Association, <www.wwindea.org>.
9 WWF, <www.panda.org> on wind power.
10 <www.acciona.com>.
11 WWF, <www.panda.org>.
12 Ibid.
13 Ibid.

14 Ibid ('hydropower, blue or gold?').
15 WWF, <www.panda.org>.
16 Ibid ('Energy solutions: Geothermal energy').
17 WWF, <www.panda.org>.
18 Ibid.
19 Ibid.
20 <www.iaea.org>.

Opposite above: Which looks more appealing, almost silent, clean wind turbines, or a large atmospheric polluting power station?

Opposite below: One standard 1.5 megawatt wind turbine can, during its operating lifetime, keep about 80,000 tons of coal out of the atmosphere. This is equivalent to about 3,000,000 tons of CO_2.

Opposite above: A solar power plant, like the one now operating in Seville, focuses the sun's energy using large mirrors in order to heat water, which in turn produces steam, which is used to power a turbine to produce zero greenhouse gas electricity.

Opposite below: The sun's energy can be used to heat water, using solar panels, and electricity using photovoltaic panels, with zero greenhouse gas emissions. These are becoming much more common on the roofs of houses and commercial premises.

S

THE SUN

The sun is a star, which is located at the centre of our solar system. It is only one of about 100 billion stars in our own galaxy, the Milky Way, which in turn is one galaxy of literally billions in the universe, each containing billions and billions of stars.

Our sun has been in existence for about 4,600,000,000 years, and it will provide life support for planet Earth for another 5–6,000,000,000 years, until it transforms into a red giant gas star and finally dies.

The energy from our sun, in the form of sunlight, supports life on Earth and is responsible for the Earth's weather and climate systems.

The sun cannot be ignored in the context of global warming.

Earth's power supply

The surface of the sun has a temperature of about 5,500°C (9,932°F), and it's even hotter on the inside! As you might expect, being the only source of heat in our solar system, its effects on Earth's climate and temperature are significant.

The total amount of radiant energy emitted by the sun that reaches Earth is termed the total solar irradiance or TSI, and it is measured in watts per metre squared.

Energy from the sun is measured both at the top of the Earth's atmosphere, called the solar constant, and at the surface of the Earth, insolation.[1]

The amount of energy measured by satellite on one square

metre of the Earth's outer atmosphere is generally accepted as being 1,368 watts per metre squared. The amount of insolation on the Earth's surface on a clear day has been measured at 1,000 watts per metre squared, but this greatly depends on atmospheric variables, such as cloud cover, etc.[2]

The calculations used by scientists to measure these figures are, as you can imagine, extremely complex. However, when it comes to climate, calculating incoming insolation is very important.

Has the sun caused or contributed to global warming?

This is the question many scientists are trying to answer. NASA satellites have measured total solar irradiance since 1978. Six overlapping satellites have monitored TSI since 1978, and the first records came from the Nimbus 7 Earth radiation budget (ERB) experiment from 1978–1993. NASA's ACRIM 1 satellite, which is an acronym for Active Cavity Radiometer Irradiance Monitors, also studied the sun, from 1980–1989, and ACRIM 2 from 1991–2001. Finally ACRIM 3 from 2000–2005.[3]

These satellites produced a wealth of information about the sun. Richard Wilson, a researcher affiliated with NASA's Goddard Institute for Space Studies and Columbia University Earth Institute in New York, compiled TSI records over the twenty-four-year observation period by piecing together the records.

The results showed a 0.05 per cent decade upward trend of TSI measured in watts per metre squared between solar minimum solar cycles, 1978 to the present (solar cycles twenty-one to twenty-three). According to Richard Wilson, this trend is important because if sustained over many decades it could cause significant climate change. We will return to this point later in the chapter.[4]

What are solar cycles?

The sun goes through cycles, called solar cycles, every eleven years. During this period the sun goes through a period of increased magnetic and sunspot activity, called the solar maximum, when solar-energy output increases, followed by a quieter period, called the solar minimum, and back again. During a solar cycle, the number of sunspots also varies with solar minimum and solar maximum, with peak sunspot activity occurring at the solar maximum.[5]

Richard Wilson's results show that the sun's TSI has increased by 0.05 per cent per decade during the sun's 'cooler' quieter periods, the solar minimum.[6]

The IPPC February 2007 report confirms that data taken over the last twenty-eight years also shows that during solar cycles, solar irradiance appears to increase by about 0.08 per cent between solar cycle minima and maxima.

What are sunspots?

Sunspots are basically dark and relatively cooler regions of the sun, caused by concentrated magnetic fields. Sunspots can cause decreases in TSI by about 0.2 per cent during say a week-long passage of a large sunspot group across the 'Earth facing' surface of the sun. These changes are insignificant, however, to the sun's total output of energy, but still equivalent to all the energy mankind produces and consumes in one year![7]

So, when the sun is at solar maximum, irradiance and magnetic activity are at their highest, which is proportional to solar activity. Sunspot numbers are representative of the general level of solar activity.

At present the sun is just coming out of a quiet period, which is solar minimum of solar cycle twenty-three. The last solar maximum was in about the years 2000–2002 (cycle twenty-three).

NASA scientists have recently discovered a new technique, 'helioseismology',[8] which works in a similar way to ultrasound, but in the case of the sun, not part of the human body!

The sun's magnetic fields, plasma flows and magnetic signatures left by fading sunspots are looked at by NASA's solar and heliosphere observation satellite (SOHO). This has led the NASA team to predict that the sun's next solar cycle will begin with an increase in solar activity in late 2007 or early

2008, and this will be thirty to fifty per cent more intense than the current cycle, reaching its peak in about 2012 (cycle twenty-four).[9]

This could affect space satellites and any technology that relies on them, as the sun's energy output increases together with perhaps a temporary increase in the Earth's temperature during this period.

Has the sun affected the Earth's climate in the past?

While NASA satellites have been monitoring sunspot activity and TSI since 1978, scientists and astronomers have been looking at the sun through telescopes for almost 400 years, since shortly after the telescope was invented.

As we know from Chapter H, the Earth entered a cooling period about 1350–1850, which was termed the Little Ice Age. Temperature drops around the globe were noticeable. Glaciers in the Alps advanced, canals in Holland regularly froze and the Thames in London would freeze over every twenty years or so.

During the coldest part of the Little Ice Age, between about 1645 and 1715, very low sunspot activity was observed.[10]

This period is called the Maunder Minimum, after the English astronomer who made the observation.

Scientists now consider there is a link between the Little Ice Age and the low level of sunspot activity recorded during that time.

So, it would seem that from scientific studies of the sun so far that the sun's irradiance may be slightly increasing by 0.05 per cent each decade, which may have an effect on climate change over timescales of 100 years or more.

The sun's energy output may be responsible for causing or contributing to the Little Ice Age, as very low sunspot activity was observed for seventy years during the coldest part of this period.

Is it the sun, yes or no?

While there is little doubt about the fact that global temperatures have increased during the last 100 years, there is continuing scientific debate in respect of the sun's contribution to current global warming.

According to the 2007 IPCC report, changes in solar irradiance since 1750 are estimated to have caused a radiative forcing of 0.12 (+0.6 to +0.30) watts per square metre.

This is compared to a total net anthropogenic (manmade) forcing of 1.6 (+0.6 to +2.4) watts per square metre.[11]

Manmade radiative forcing is therefore much greater than the effect the sun has had warming the Earth since the year 1750.

Radiative forcing is basically the change in the balance between radiation entering the Earth's atmosphere and leaving it. Positive forcing will warm the Earth and negative will cool it.

It seems therefore that while the sun does of course have an effect on the Earth's climate, and therefore potentially global warming, such effects are nowhere near as great as those of anthropogenic or manmade causes, the burning of fossil fuels, etc. However, much longer studies will have to be made, it seems, to determine the answer for sure!

Having considered the sun's role, we will now look at one of the most serious consequences of global warming, Earth's rising temperature.

Key points

- ➢ Light energy emitted from the sun, solar irradiance, has increased by about 0.05 per cent each decade from 1978, which could affect Earth's climate over the long term.
- ➢ The sun goes through solar cycles, which occur every eleven years.
- ➢ Sunspot numbers represent the general level of solar activity, with peak sunspot activity occurring during solar maximum.
- ➢ During the coldest part of the Little Ice Age, sunspot activity was observed to be at its lowest.

➢ While the sun does of course have an effect on Earth's climate, solar irradiance has had a radiative forcing of 0.12 watts per metre squared on Earth's atmosphere, compared with an anthropogenic forcing of 1.6.

---

1 NASA, <www.nasa.gov/goddard> ('Solar radiation and the Earth system').
2 NASA, <www.nasa.gov/goddard>.
3 Ibid.
4 NASA Astrobiology Institute, <www.nai.nasa.gov> ('Solar influence', Part 3).
5 NASA, <www.nasa.gov>.
6 Ibid.
7 Ibid.
8 NASA, <www.nasa.gov> (NASA aids in resolving long-standing solar-cycle mystery, March 2006 study).
9 NASA, <www.nasa.gov>.
10 NASA Astrobiology Institute, <www.nai.nasa.gov> ('Solar influence', Part 3).
11 IPCC Climate Change 2007, 'The physical science basis'.

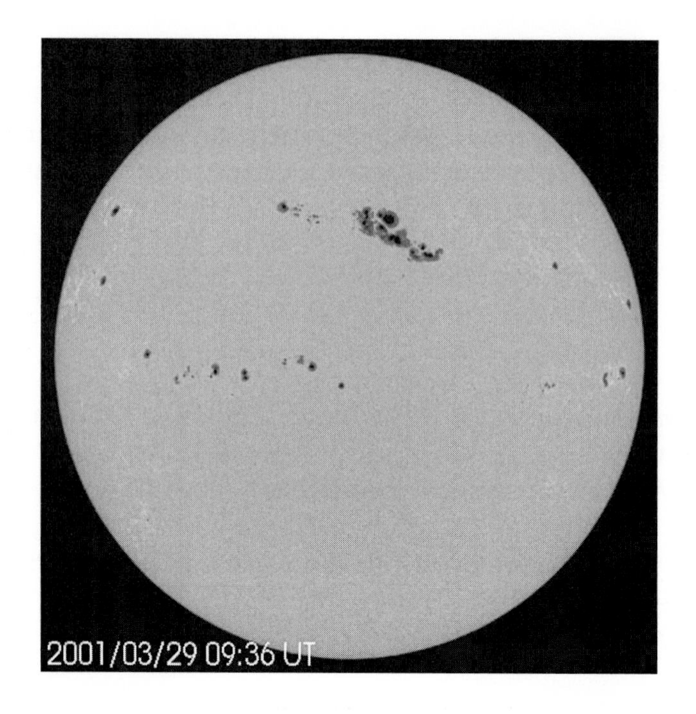

Above: Active region 9393 as seen by the Michelson Doppler Imager (MDI) instrument on SOHO hosted the largest sunspot group observed so far during the current solar cycle. On 30th March 2001 the sunspot area within the group spanned an area more than thirteen times the entire surface of the Earth!

Credit NASA <www.nasa.gov>.

Opposite above: ACRIM composite Total Solar Irradiance graph.

Credit NASA, <www.nasa.gov>.

Opposite below: ACRIM composite Total Solar Irradiance and Greenwich sunspot number graphs.

Credit NASA, <www.nasa.gov>.

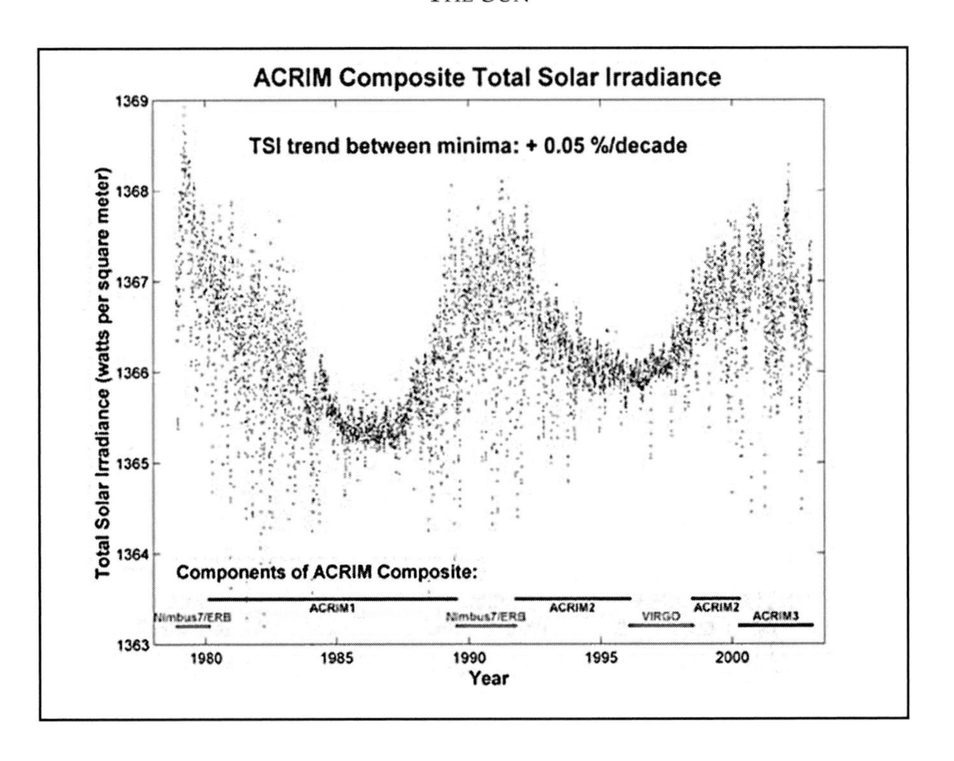
ACRIM Composite Total Solar Irradiance
TSI trend between minima: + 0.05 %/decade
Total Solar Irradiance (watts per square meter)
1369
1368
1367
1366
1365
1364
1363
Components of ACRIM Composite:
Nimbus7/ERB
ACRIM1
Nimbus7/ERB
ACRIM2
VIRGO
ACRIM2
ACRIM3
1980
1985
1990
1995
2000
Year

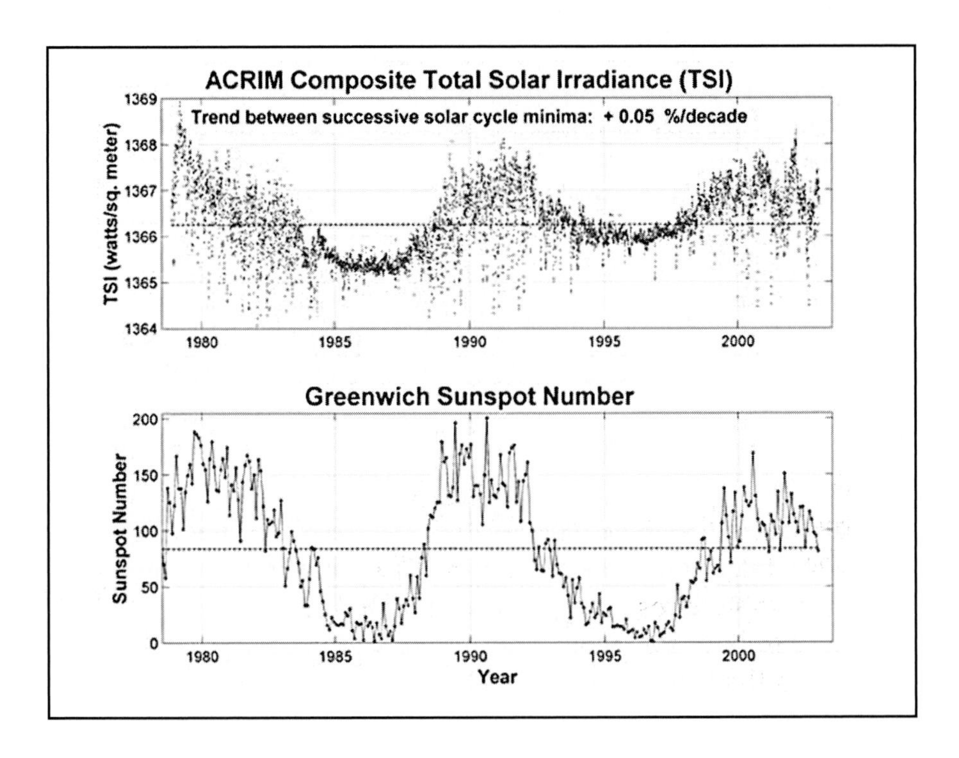
ACRIM Composite Total Solar Irradiance (TSI)
Trend between successive solar cycle minima: + 0.05 %/decade
TSI (watts/sq. meter)
1369
1368
1367
1366
1365
1364
1980
1985
1990
1995
2000
Greenwich Sunspot Number
Sunspot Number
200
150
100
50
0
1980
1985
1990
1995
2000
Year

T

TEMPERATURE

We now arrive at Chapter T – T for Temperature in the A–Z of global warming, the increase of which is the main consequence of the higher levels of greenhouse gases within Earth's atmosphere.

Temperature is generally measured using the Celsius scale, except in the USA, where the Fahrenheit scale is used. Zero degrees Centigrade corresponds to the temperature at which water freezes, and 100 degrees when it boils. These temperatures are represented as 32 and 212°F respectively.

The Earth's average temperature, assisted by its naturally occurring greenhouse-gas blanket, is about 15°C (59°F). The average temperature of the human body is about 37°C (98°F), and if temperatures get too high harmful reactions and even death may result.

Just like a human being, if Earth's temperature increases too much, the planet will start to get sick and serious consequences will result, some of which are already becoming evident.

How much has the Earth's temperature increased?

The Earth's global mean surface temperature according the Fourth Assessment Report of the IPCC puts the rise at 0.74°C (1.33°F) over the period 1906 to 2006.[1]

Global temperature is measured by taking the average near-surface temperatures over air, sea and land.

This rise may not seem like much, but according to NASA, this means that the Earth is now reaching and passing through the warmest period in the current interglacial period, which has lasted for nearly 12,000 years.[2]

How fast is Earth's temperature rising?

The Earth's temperature has risen by about 0.2°C (0.36°F) each decade over the last thirty years. The studies show that warming is greatest at higher latitudes of the northern hemisphere, and larger over land compared to the oceans, as the oceans have a much higher heat capacity compared to the land.[3] Air temperatures in the Arctic region for example have, on average, actually increased by about 5°C (9°F) over the last 100 years.

What about historical warming?

We know from Chapter H that the Earth has had many periods of warming and cooling, and historically these temperature changes have had little to do with manmade greenhouse gases, as mankind has been emitting greenhouse gases significantly only since the Industrial Revolution, in and about the late nineteenth century.

Two of the most recent temperature changes took place during the Little Ice Age, in the years 1350–1850, or thereabouts, when temperatures dipped, and the Medieval Warm Period between years 1000–1300, or thereabouts, when temperatures got comparatively warmer again. An explanation for the Little Ice Age, or Maunder Minimum is the lack of sunspot activity and solar irradiance that occurred during this time (as explained in Chapter S).

What about more recently?

Well, temperatures have been measured accurately with scientific instruments for about only 150 years or so. Prior to this a range of proxy data is used, such as tree rings, ice cores, lake and sea sediments, corals and historical records.[4]

Researchers from NASA, Dr James Henson and his colleague Mark Imhoff, analysed records from 7,500 global weather stations and used satellite observations of night-time weather stations to identify minimal human influence, such as urban heat island effects.

The team concluded that from 1900 to 1940 it was possible the Earth warmed partly as a result of increased levels of greenhouse gases and partly due to natural climate variability.

Between 1940 and 1965 the Earth cooled by about 0.1°C (0.18°F), which some scientists attribute to the increased use of aerosols and other airborne pollutants from the burning of fossil fuels. This was especially so in the northern hemisphere, where cooling occurred most during this period, which can lead to increased cloud cover, which in turn blocks and reflects incoming solar radiation. This is a phenomenon that has been termed 'global dimming'. Aerosols, certainly in the northern hemisphere, have been slowly phased out, however, which may have helped reveal the true extent of greenhouse-gas-induced warming.

The period from 1965 to 2000 showed large and widespread warming around the world.[5]

Indeed the IPCC concluded in 2001 that there is new and stronger evidence that most of the warming observed at least over the past fifty years is attributable to human activities.

Link between global warming and human activities?

There has been much debate between scientists over attribution of climate change and global warming, and much of this discussion has focused on a temperature graph produced in 1999 for the IPCC, by climatologist Michael Mann and his colleagues, which showed temperatures extending back 1,000 years. The debate became known as the 'hockey stick' debate.

This name came from the graph itself, as it shows temperatures for about 1,000 years remaining more or less constant, then from about 1800 a sharp upward trend occurs that resembles the end of a hockey stick.[6] The reconstructions showed the 1990s to be the warmest decade, with 1998 the warmest year ever.

The graph seems to support the warming influence human beings have had on climate over the last 150 years or so, as evidenced by the sudden upward trend in temperatures recorded.

Certain criticism was made of the fact that accurate temperature records go back only 150 years, and that the data and methods used to recreate the temperature prior to about 1850 cannot be reliable as it comes from proxy sources such as tree rings, corals and ice cores, etc.

It would appear, however, that much of the debate as to who is responsible for global warming is now settled. While solar intensity and even volcanoes and other natural factors can explain variations in global temperatures in the early nineteenth century, rising greenhouse gas levels can provide the only plausible explanation for the warming trend over the past fifty years.[7]

In response to the controversy over the Mann temperature graph, in 2006 the US Congress requested the National Research Council prepare a report. They concluded that there was a high level of confidence that the global mean surface temperature during the past few decades is higher now than at any time over the preceding 400 years. There is less confidence prior to the year 1600 to support temperature reconstructions, as there is less data available from whatever source. There was even less confidence about the conclusions reached that the 1990s were the warmest decade and 1998 the warmest year. The committee did indicate, however, that none of the reconstructions showed that temperatures were warmer during medieval times than during the last few decades.[8]

The main conclusion, however, is that the build-up of greenhouse gases in the atmosphere will cause several degrees of warming, and this is based on the laws of physics and chemistry. The link between greenhouse gases and temperature is well established, as we know from Chapter G, so when additional CO_2 is added to the atmosphere, by burning fossil fuels, the temperature is going to increase. This has been confirmed by reliable scientific instruments over the last 150 years.

How high will temperatures go?

For the last three decades temperatures have been rising by about 0.2°C (0.36°F) per decade. There is evidence however that the warming may accelerate as positive feedback mechanisms come into play. Examples would be the release of methane from the ground as the permafrost starts to melt, thus accelerating the warming. Studies already indicate that warming is greater over the northern hemisphere. As the snow and ice melt in the Arctic regions, darker surfaces are uncovered, which reduces the albedo effect of the ice/snow-covered areas, which allows more sunlight to be absorbed, thus increasing warming. Likewise as the atmosphere warms it is able to hold more water vapour (itself a greenhouse gas), which allows it to trap more heat. These are two examples of positive feedback mechanisms.[9]

It is not yet possible, however, to determine what temperature will result from a certain level of greenhouse gas.

It is estimated that if greenhouse gas could be stabilised at today's level of about 430 ppm CO_2 equivalent, the Earth would be committed to an eventual temperature increase of about 1–3°C (1.8–5.4°F) above pre-industrial levels.[10]

Projected CO_2/temperature level scenarios

The amount the Earth's temperature goes up depends on greenhouse gas levels in the atmosphere.

Projections of future warming depend on projections of global emissions. If emissions were to remain at today's levels, then greenhouse gas would reach about 550 ppm CO_2e by about 2050, based on the current annual increase of 2.5 ppmv CO_2e. This would commit the world to a temperature rise of about 2–5°C (3.6–9°F).[11]

The IPCC however projects that without intervention greenhouse gas levels will rise to 550–700 ppm CO_2e by 2050, and 650–1,200 ppm CO_2e by 2100! This would cause temperature rises of between 1.5–4.4°C (2.7–7.9°F) and 1.8–5.5°C (3.2–9.9°F) respectively, just on the lower forecasts of 550 and 650 ppm CO_2e levels alone![12]

> 'A temperature rise of 2–3°C (3.6–5.4°F) above present levels would put the Earth at a temperature not experienced for three million years and far outside the experience of human civilisation.'[13]

The Earth is already committed to a 1–3°C rise (1.8–5.4°F) on current greenhouse gas levels. If the Earth warms by a further 1°C (1.8°F), NASA scientists point out that this will be the warmest Earth has been for the past 1,000,000 years. At 2 or 3°C higher (3.6–5.5°F), the Earth would become a different world from that we know. As mentioned above, the last time this occurred was about 3,000,000 years ago, and sea levels are estimated to have been twenty-five metres higher (eighty feet) than present![14]

There seems to be no alternative, therefore, other than for humankind to reduce greenhouse gas emissions, significantly, and fast, in order to prevent disastrous consequences. The big problem is that like a huge oil tanker trying to make a U-turn, even if emissions could be halted now, the effects of current levels will continue to cause temperatures to rise for a long time to come.

Any evidence of increasing temperatures currently affecting Earth?

According to the WWF, evidence comes from the bleaching and degradation of coral reefs (discussed further in Chapter V), due to increasing sea temperatures, which could degrade Australia's Great Barrier Reef in a single human lifetime. Alpine forests struggle to spread to higher, cooler locations, and glaciers are melting all over the world.[15]

The Caribbean saw its warmest ever ocean temperature in 2005.

Scotland in the UK saw its hottest year on record in 2003, which caused hundreds of adult salmon to die, as the water became too warm for the fish to extract oxygen from it.

New modelling work by the UK's Hadley Centre shows that the summer of 2003 was Europe's hottest for 500 years.

In the Arctic, sea ice measurements in 2007 recorded the smallest sea-ice cover ever at the end of the summer melt season.

In 2003, the world's major cities sweltered under heatwaves. In France, during the summer of 2003, the heatwave killed about 14,800 people in Paris alone, according to official figures released in September 2003.

Summer temperatures have been analysed in sixteen of

Europe's cities, which show that the continents' capitals have warmed by up to 2°C (3.6°F) in the last thirty years.

London is the city where average maximum summer temperatures increased the most, up 2°C (3.6°F) over the last thirty years, followed by Athens and Lisbon (1.9°C or 3.4°F), Warsaw (1.3°C or 2.3°F) and Berlin (1.2°C or 2.1°F).[16]

Between 2000 and 2005, average summer temperatures in thirteen out of sixteen cities looked at were at least 1°C (1.8°F) higher than during the period 1970–1975.

Earth's warmest years

According to climatologists at NASA's Goddard Institute for Space Studies the five warmest years since the 1880s have been

1 2005
2 1998
3 2002
4 2003
5 2006

The year 2005 therefore has been the hottest so far, though it shares this accolade with 1998, which was virtually as hot. Year 1998 temperatures were enhanced, however, by the strongest tropical El Niño for almost a century, which boosted temperatures above the level they otherwise would have been. As the El Niño gets underway in the topical Pacific Ocean, 2007 could be even hotter, bringing with it increased warmth. El Niño is discussed further in Chapter W.[17] Since editing, NASA has announced that 2007 has tied with 1998 as the second warmest year in a century, while 2005 maintains its first place for the moment!

A 2°C (3.6°F) increase limit

The WWF is advocating that temperatures cannot be allowed to rise by more than 2°C (3.6°F) above pre-industrial levels, otherwise dangerous climate change may occur.[18] The Earth has already warmed by 0.74°C (1.33°F), which means another 1.3°C rise (2.34°F) could be too much.

The 2°C (3.6°F) threshold is based on the best available science and is accepted by many governments including the prime ministers and presidents of all twenty-five EU member states.[19]

The only way to prevent temperatures staying below this level is for CO_2 concentrations to stay below about 400 ppmv, the equivalent to greenhouse gas levels of around 450co_2e. If this were possible, staying below 2°C (3.6°F) is likely, according to climate models.[20] Levels of CO_2, however, are already at 385 ppmv, which means the chance of stabilisation below 400 ppmv is therefore very unlikely.

What would a 2°C (3.6°F) rise in temperature mean?

The WWF has looked at three regions to see what a 2°C (3.6°F) temperature rise would mean for those regions.

The Mediterranean

Everyone enjoys going on holiday to the 'Med', with its beautiful warm climate. However, as temperatures rise in the region, water shortages could become common as annual rainfall could decrease by twenty per cent, and more heatwaves cause all-year-round risk from serious forest fires, as maximum temperatures could rise by up to 5°C (9°F).

The Arctic

Temperatures would rise by about 3.2°C (5.7°F) here, maybe even double that if temperatures rose by 2°C (3.6°F) elsewhere. Less ice means more heat absorption as the darker water absorbs the sun's energy. Arctic summer ice could totally disappear, leaving wildlife habitats, such as the polar bear's, deteriorating or destroyed.

Eastern Canada

Important species of trees, including the sugar maple, Canada's national symbol, will be forced to move northwards, which could cause problems if the trees cannot adapt. Canadian fisheries will also struggle, which could be the final straw for the already endangered Atlantic salmon.[21]

These are just examples of three regions and the effects of a 2°C (3.6°F) rise in temperature. Of course, many other regions would also suffer similar consequences.

According to the Stern Review on the Economics of Climate Change, some climate models suggest that a global 2°C (3.6°F) rise above pre-industrial levels would mean that there is potential for the Greenland ice sheet to begin melting irreversibly, a rising risk of the collapse of the West Antarctic ice sheet, and a rising risk of the collapse of the ocean thermohaline circulation.[22]

If temperatures rose more than 5°C (9°F), which is possible if emissions continue to grow, and positive feedback mechanisms kick in, such as the release of CO_2 from carbon sinks and methane from permafrost, then the rise in temperatures would be equivalent to the amount of warming that took place between the end of the last Ice Age and today.[23]

Such a rise in temperature would be far outside human experience. A very sobering thought!

The Earth, like a sick human being, is already beginning to show the effects of higher temperatures. A 2°C (3.6°F) global temperature rise appears to be the limit recognised as causing catastrophic climate change.

Staying below 2°C (3.6°F) requires CO_2 levels to be stabilised at 400 ppmv, and this appears unlikely as CO_2 levels are already at 385 ppmv and increasing annually. Greenhouse gas levels are already at 430 ppm CO_2e, and rising at 2.5 ppm CO_2e annually. If this continues, the Earth may well be 2–5°C (3.6–9°F) warmer by 2050, when greenhouse gas levels would reach about 550 ppm CO_2e.

It seems the only answer will be for all nations and all individuals to do their bit as far as possible to prevent, or at least reduce, greenhouse gas emissions. The science appears clear. While it may not be possible to prevent a 2°C (3.6°F) temperature rise, it seems everything must be done to prevent rises over and above this level, and the window of opportunity to do so is rapidly disappearing.

In the eleven months it has taken me to write this book, CO_2 levels will have risen by about 1.28 ppm and greenhouse gas levels generally by about 2.29 ppm CO_2e.

We now reach the end of the twentieth chapter (of twenty-six), and up to now the book has dealt mainly with the scientific facts behind global warming and its causes. The remaining chapters look at what the effects will be on Earth's inhabitants and the weather, and also at what can be done to try and prevent, or at least slow down, one of the biggest threats to planet Earth. First though, for anyone who may not yet be convinced about global warming, the next chapter is for you! It deals with some of the arguments being put forward by the global warming sceptics.

Let's look at some of the arguments in Chapter U (for Unconvinced).

Key points

- ➢ Earth's global mean surface temperature has increased by 0.74°C (1.33°F) over a hundred-year period, 1906–2006.
- ➢ Temperatures in the Arctic, however, have increased by about 5°C (9°F) over a similar period.
- ➢ If greenhouse gases could be halted at present levels, the Earth would still warm by about 1–3°C (1.8–5.4°F) above pre-industrial levels (possibly 2.26°C more than present).
- ➢ The last time Earth was 2–3°C (3.6–5.5°F) higher than present was 3,000,000 years ago, when sea levels may have been twenty-five metres (eighty feet) higher than present.
- ➢ The warmest year since 1880 was 2005, virtually on a par with 1998, when temperatures were boosted by an exceptional El Niño year, while 2007 has become Earth's second warmest year jointly with 1998.

1. IPCC Climate Change 2007, 'The physical science basis'.
2. NASA, <www.nasa.giss.nasa.gov>.
3. Ibid.
4. Mongabay, <www.mongabay.com> (Earth at its warmest for 400 years/National Academy of Science, 22nd June 2006).
5. NASA, <www.nasa.giss.nasa.gov> (research news, 5th November 2007).
6. Stern Review on The Economics of Climate Change, Part I.
7. Ibid.
8. Mongabay, <www.mongabay.com>.
9. Stern Review on The Economics of Climate Change, Part I.
10. Ibid.
11. Ibid.
12. Stern Review on The Economics of Climate Change, Part I/IPCC TAR.
13. Stern Review on The Economics of Climate Change, Part I.
14. NASA, <www.nasa.giss.nasa.gov> (research news, 5th November 2007).
15. WWF, <www.panda.org>.
16. WWF, <www.panda.org> ('Europe feels the heat').
17. NASA Godard Institute for Space Studies, <www.giss.nasa.gov>.
18. WWF, <www.panda.org>.
19. WWF, <www.panda.org> ('2 degrees C is too much').
20. WWF, <www.panda.org>.
21. WWF, <www.panda.org> ('2 degree scenarios').
22. Stern Review on The Economics of Climate Change, Part I.
23. Ibid.

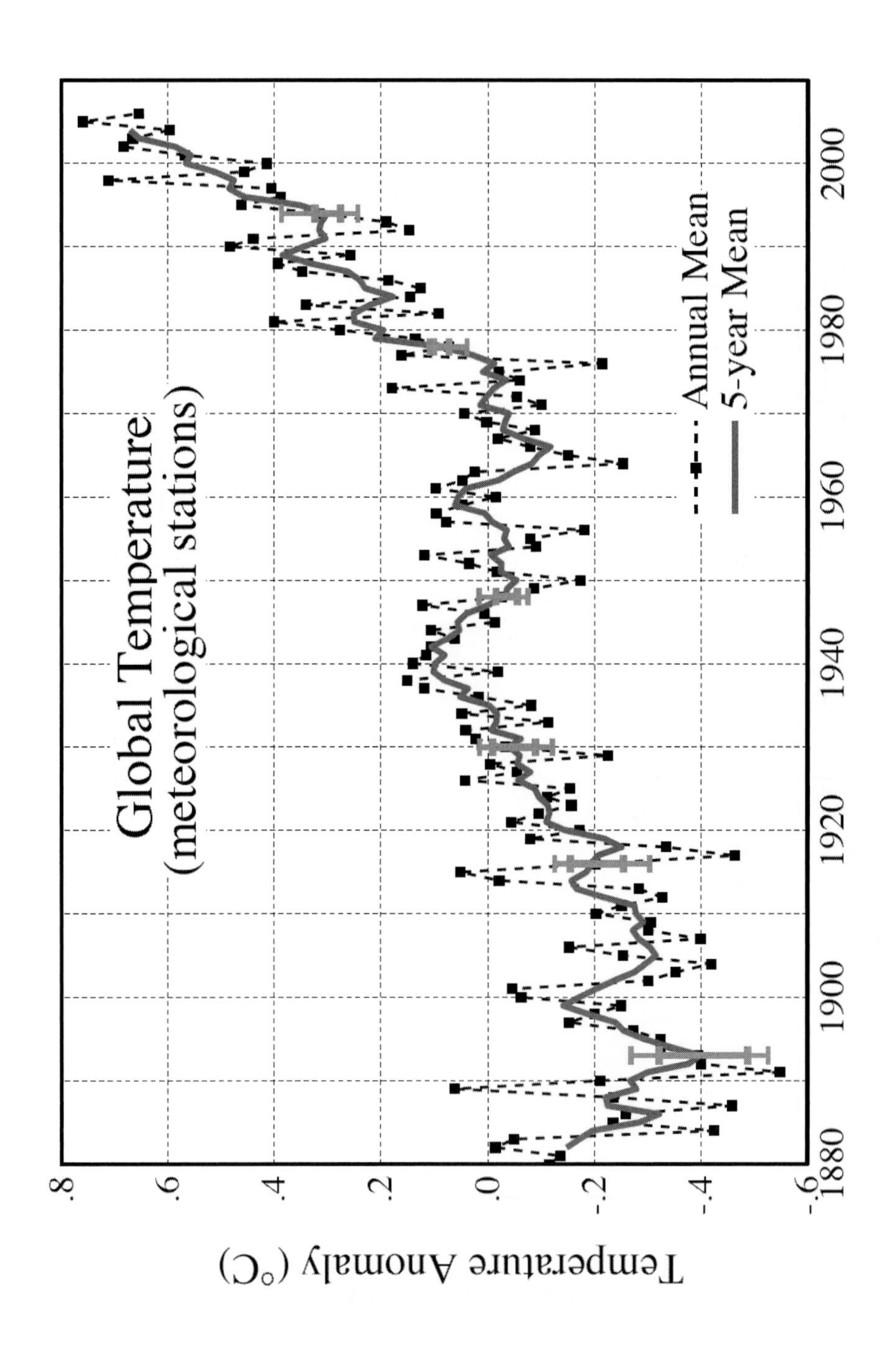

Credit NASA, <www.giss.nasa.gov>.

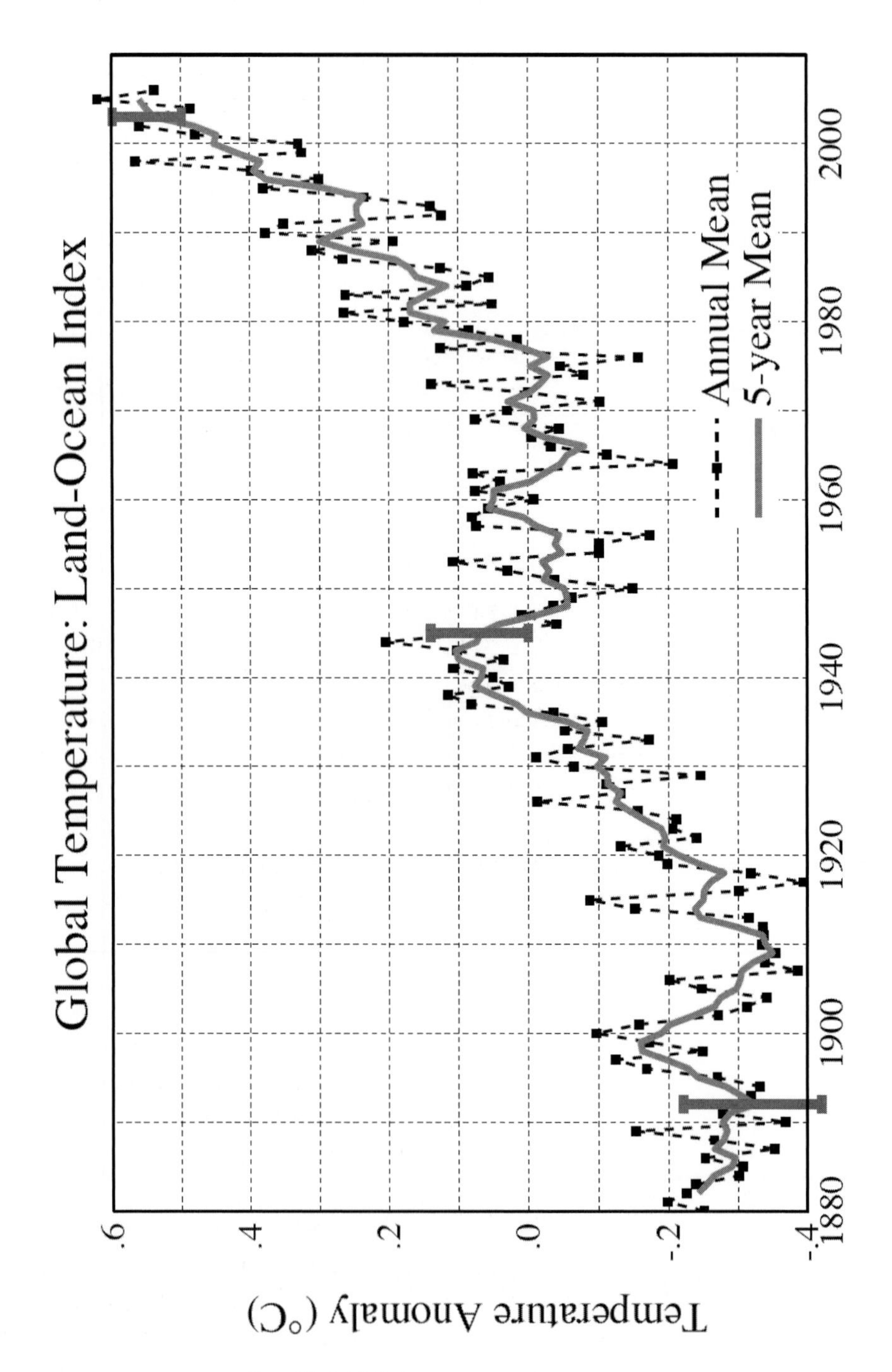

Credit NASA, <www.giss.nasa.gov>.

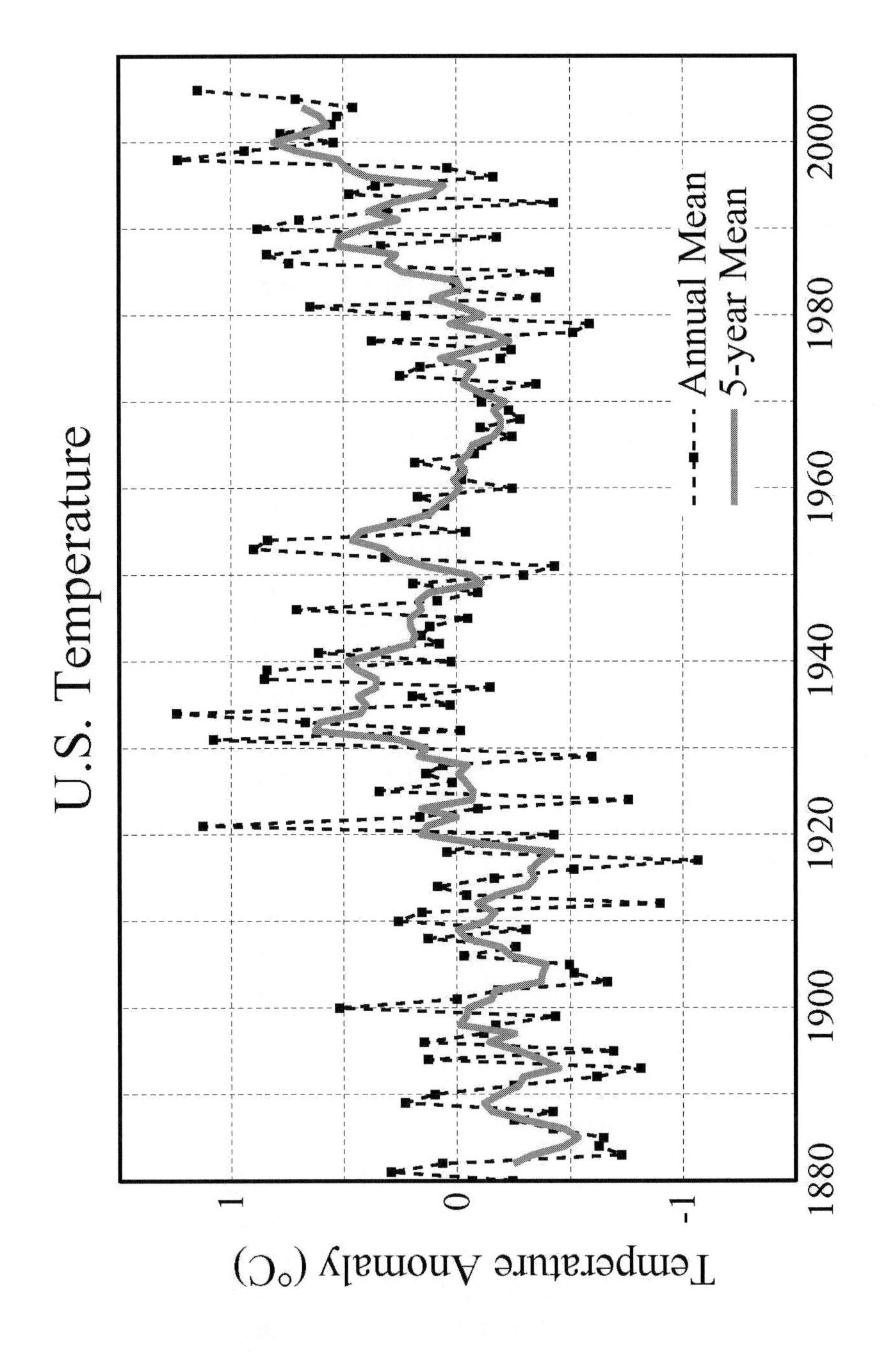

Credit NASA, <www.giss.nasa.gov>.

U

UNCONVINCED?

We have now reached a point in the book where we will revisit and clarify a few matters in this rather complex subject. We will also consider some of the main points usually cited by global-warming sceptics to support their argument that global warming is not really happening, or at least that it's not our fault!

Global-warming theory does of course have its share of sceptics – some scientists, some not – and it's easy to understand why. Sometimes the facts appear confusing. For example, during the first four decades of the century, temperatures rose, suggesting a warming trend. Then from 1940–1965 they dropped again, which no doubt prompted some scientists at the time to warn of an impending ice age!

In this chapter we will look at some of the arguments raised by the sceptics and the responses, most of which have already been dealt with to some extent in the preceding chapters, in an attempt to correct any confusion about the issues.

Okay, let's look at some of the points raised about the main culprit, $CO_{2.}$

Natural CO_2 emissions are far greater than humankind's

It is true to say that emissions of CO_2 from natural sources, such as the breakdown of organic matter, respiration of living organisms etc., are far greater than humankind's CO_2 emissions. Prior to the Industrial Revolution, however, usually taken to be about the year

1750, natural emissions of CO_2 were balanced by the absorption of CO_2 by natural sinks, such as the land, soil and the oceans. Since humankind started burning fossil fuels, the balance has been disrupted, and now about twenty-six gigatonnes (26,000,000,000 tonnes) of CO_2 are emitted annually, with fifty per cent of it being absorbed by the natural sinks, and the remaining fifty per cent staying in the atmosphere. It's the additional CO_2 that is increasing the greenhouse effect and contributing to the warming up of the Earth.

Evidence also comes from the fact that CO_2 trapped in ice cores shows fairly consistent levels of CO_2 in the atmosphere of between 180 to 300 ppm over the last 650,000 years, whereas only in the last 150 years have CO_2 levels climbed to their present level of about 385 ppm, which corresponds to humans first starting to burn fossil fuels.

More CO_2 comes from volcanoes

This is simply incorrect. CO_2 generated from volcanic eruptions is nowhere near the annual twenty-six gigatonnes emitted from humans worldwide. If this statement were correct, then CO_2 levels measured at Mauna Loa in Hawaii since 1958 would show sudden fluctuations after each significant volcanic eruption. The Keeling curve, which shows a steady rise in CO_2 levels since 1958, shows no such variations.

Furthermore NASA researchers at the Lawrence Livermore National Laboratory, after examining data between 1979 and 1999, have discovered that large volcanic eruptions actually cooled the lower troposphere (first 8 kilometres, or 4.9 miles, of Earth's surface). They found that aerosol particles from one of the largest recent eruptions of Mount Pinatubo in the Philippines in 1991 cooled the lower atmosphere and probably actually masked the real warming of the atmosphere for a short while after the eruption.[1]

CO_2 measurements from Mauna Loa must be flawed as they are taken from an active volcano!

It is true to say that Mauna Loa is the world's largest active volcano. The location chosen by Professor Keeling and his team is actually a good place to take measurements from, as Mauna Loa is situated on Hawaii, which is in the middle of the Pacific and away from major industrial pollution sources. Professor Keeling would have course known that Mauna Loa was an active volcano, and this fact would have been taken into account when CO_2 measurements were taken. The Keeling graph produced from measurements obtained shows a clear saw-toothed graph, which represents the fluctuating amounts of CO_2 in the atmosphere, rising and falling with the seasons.

Many other CO_2 recording stations based all around the world, including one based in the Antarctic all show similar results to the Keeling graph from Mauna Loa, thus confirming the Mauna Loa results.

CO_2 is not responsible for warming as temperatures dropped after 1945 through to 1965 just as increased industrialisation accelerated after World War II

According to NASA the Earth cooled by about 0.1°C or 0.18°F between 1945 and 1965. It appears widely accepted by the scientific community that this was not because CO_2 levels were not affecting temperature, but because increased use of aerosols during this time (which create fine particles in the air) tend to block incoming solar radiation. This in turn may have masked the true warming caused by rising CO_2 levels.[2]

As aerosol use was phased out during the 1970s, CO_2 warming that had been masked revealed itself once again.

What about temperature?

It's been warmer before, therefore warming is part of a natural cycle

This is very true, see Chapter H. However, there is clear evidence that shows accurate temperature measurements have been taken only for the last 150 years. The five warmest years since the late 1880s, according to NASA, have been 2005, 1998, 2000, 2003 and 2006, with 2005 the hottest so far.

The year 2005 is probably ranked hottest despite 1998 being on a par, due to the fact that 2005 temperatures were not boosted by a tropical El Niño.

The year 2007 may turn out to be the warmest in the period of instrumental measurements, according to James Henson, director of the NASA Goddard Institute for Space Studies.[3] It's difficult to believe while writing this section, during a windy, cold and wet July evening in the UK, but no doubt all will be revealed in 2008!

Having said that, on editing the book in February 2008, NASA's Goddard Institute for Space Studies has just announced that 2007 now ranks as Earth's second warmest year this century, jointly with 1998! 2005 keeps hold of the record so far.

It is true to say, however, that the Earth is now reaching temperatures not experienced for about 12,000 years.

The Earth has been both colder, during the ice ages brought about by the Earth's changing orbit around the sun, and warmer, during events like the Paleocene Eocene Thermal Maximum (PETM), brought about, it is thought, by massive volcanic eruptions. However, manmade greenhouse gases now appear to be pushing temperatures up far beyond the natural cycle of the last 12,000 years, which is something everyone should be concerned about.

It's the sun causing rising temperatures, not CO_2!

The sun's role in global warming has already been considered in Chapter S. However, for the sake of completeness, it has been estimated that the sun's irradiance has been increasing by a very small 0.05 per cent per decade. Such an increase may be able to influence climate change if maintained over 100 years or so. The sun's irradiance has been monitored only since 1978, and therefore records do not really go back far enough to determine whether the sun has had a long-term effect on climate over the last 150 years or so.

The above fact, however, does not seem to provide the reasons for the fastest observed decadal rise in temperatures (0.2°C or 0.36°F) globally during the last thirty years.

Also, it has been estimated that the sun has a radiative forcing of about 0.12 watts per square metre compared to humankind's atmospheric radiative forcing of 1.6 watts per square metre.

A hypothesis put forward by Danish scientists suggests that the sun's cosmic rays may affect Earth's temperature. The suggestion is that higher sunspot activity causes fewer cosmic rays due to the fact that they are deflected away from the Earth by the sun's magnetic field. Fewer cosmic rays, Danish scientists say, mean less cloud cover and hence a warmer Earth. It seems, however, that this hypothesis is far from proven and further research needs to be done!

What about arguments raised in relation to ice and Antarctica?

Antarctica is really growing not melting

Studies have revealed that the centre of the Antarctic continent has been cooling, and greater precipitation (snowfall) means that the area can't be warming as it should be if global warming is really happening. Well, global-warming theory does not predict uniform warming in all places at once. We know that the edges of the continent do appear to be affected by global warming, as the West Antarctic ice sheet has started to break up, as discussed in Chapter N.

A team of NASA scientists reported in May 2007 clear evidence that extensive areas of snow melted in Western Antarctica in January 2005 in response to warm temperatures (2005 was the warmest year since the 1880s).

Melting occurred in multiple regions, including far inland, up to 900 kilometres or 500 miles from the open ocean, at high altitudes and elevations where melting had been considered unlikely. Temperatures were recorded as high as 5°C (41°F). According to Konrad Steffen, one of the NASA scientist team leaders,

> 'Antarctica has shown little to no warming in the recent past, with the exception of the Antarctic peninsula, but now large regions are showing the first impacts of warming as interpreted by this satellite analysis.'[4]

NASA also discovered that Antarctica's ice sheet mass decreased significantly between 2002 and 2005, losing about 152 (+ or – 80) cubic kilometres (36.5 + or – 19 cubic miles) of ice annually during this period. That's an awful lot of water being added to the sea when you consider that a cubic kilometre (0.6 cubic miles) is equal to a trillion litres or 264 billion gallons! Enough water to supply US consumption for three months.

Finally, what about water vapour?

Water vapour is the main greenhouse gas, and is responsible for global warming, not manmade greenhouse gases/CO_2.

Water vapour is indeed the strongest greenhouse gas, and accounts for the largest percentage of the greenhouse effect, simply because it is present in the atmosphere in greater concentrations compared to the other greenhouse gases. Water vapour is also about 99.99 per cent naturally occurring, therefore humans do not have much control over the amounts in the atmosphere in any event.

The important point about water vapour is the fact that it acts as a feedback mechanism rather than a forcing mechanism, unlike all other anthropogenic greenhouse gases in the atmosphere. This means that, as expected from basic physics, a warmer atmosphere will hold more water vapour, which traps more heat, which then amplifies the initial warming.[5] Therefore as temperatures rise, so will water vapour content in the atmosphere, which will trap more heat, and so on, until equilibrium is reached.

Water vapour acts as a strong positive feedback on global warming from the effects of the other greenhouse gases in the atmosphere.

While it is accurate to say that water vapour accounts for the strongest greenhouse gas, because water vapour accounts for the greatest gas in the atmosphere, at about four per cent by volume, it is a natural gas. It responds by amplifying the effects of the manmade greenhouse gases emitted into the atmosphere, which in turn make up a very small percentage of gases (by comparison with water vapour) in the atmosphere.

Hopefully, the above points clarify and also provide some answers to any remaining sceptics who still question the reality and causes of global warming.

We now look at what effects a warmer world may have on the transmission and prevalence of disease in Chapter V, Vectors for Disease.

Key points

- While natural CO_2 emissions are greater than those made by humans, CO_2 levels have increased from 280 to 385 ppm since the Industrial Revolution and are now higher than at any time in the last 650,000 years.
- It is thought that aerosols that were used from 1940 until the early 1960s may have reflected incoming solar radiation, cooling the Earth during this period, effectively temporarily masking true warming.
- Accurate global temperature records have been available only for 150 years, and these reveal that the last decade has seen six of the hottest years since the 1880s.

➢ Water vapour is the strongest greenhouse gas, but it is a naturally occurring gas. However, it has a positive feedback on global warming as a warmer atmosphere holds more water vapour, which in turn traps more heat and causes further warming and so on.

---

1 NASA, <www.earthobservatory.nasa.gov> (November 2003 article).
2 NASA, <www.giss.nasa.gov> (research news, 5th November 2001).
3 NASA, <www.giss.nasa.gov> (research news, February 2007).
4 NASA, <www.nasa.gov> (NASA finds vast areas of Western Antarctica melted in recent past, 15th May 2007).
5 Stern Review on The Economics of Climate Change.

V

VECTORS FOR DISEASE

Pathogens, such as viruses, bacteria and protozoans responsible for many diseases can be passed onto humans by an insect vector. A particularly important insect in this regard is the mosquito, which becomes infected when it feeds. The pathogen then reproduces and thrives inside the mosquito's body, ready to transmit whichever pathogen it is carrying onto a human host when the insect decides to bite or feast on his or her blood!

The main vector for disease that will be affected by global warming is the mosquito, which is responsible for spreading one of the most common infectious diseases, malaria. The mosquito also spreads other diseases, however, such as yellow and dengue fever. It is only the female of the species *Anopholes* that is responsible for transmitting these diseases, as it is only the female mosquito that feeds on blood, which is when transmission occurs.

Temperature and precipitation play a major role in the prevalence and abundance of vector insects such as the mosquito. Mosquitoes need stagnant water to breed, and in warmer conditions the pathogen matures more quickly inside its vector host, and the vector itself is also able to breed at a faster rate.

Other diseases that can be influenced by climate change are diarrhoeal, which usually peak during wet seasons in the tropics as water supplies become contaminated, causing others such as typhoid, cholera, E.coli, and hepatitis A, to name but a few![1]

Diseases spread by rodents, such as rats, especially after mild wet winters, which may become more common, include haemorrhagic diseases and leptospirosis, and also tick-related diseases. It is possible that the Black Death, caused by the bubonic plague in the mid-fourteenth century, may have been assisted by the warmer temperatures that prevailed during the Medieval Warm Period.

The most important vector for disease, however, is the mosquito, which can spread malaria, yellow and dengue fever, and which is currently responsible for about 1,100,000 deaths a year from malaria alone, according to the World Health Organisation (WHO).[2]

Mosquito diseases

Malaria is caused by a protozoa parasite, which is transmitted to humans by the female mosquito, which acts as the vector for the disease. Symptoms include fever, nausea, a flu-like condition, and, in severe cases, death. There is no vaccine for malaria, but preventative drugs are available to reduce the risk of infection, and nets, repellents, etc. can be used to reduce the chances of being bitten, when in mosquito-endemic areas, such as the tropics, Asia, and Africa.

Dengue fever, again found in the tropics, is transmitted to humans by the mosquito vector. Symptoms include severe headache, muscle pains and fever. A sufferer will also get a bright red rash on the lower limbs, and sometimes over other parts of the body too. In most cases the fever will last about one week and a full recovery will be made, unless you are very unfortunate and go on to develop dengue hemorrhagic fever (DHF), which may lead to dengue shock syndrome and death!

Yellow fever, another viral disease also spread by mosquitoes, is usually severe, and the virus will travel into all the organs of the body, which will cause symptoms of fever, muscle pain, shivers and vomiting. A second phase will cause haemorrhaging from the nose, mouth and eyes, and possible kidney failure. About fifty per cent of sufferers then die, but fifty per cent will go on to make a full recovery. Fortunately a vaccine for this nasty disease was developed a long time ago!

More mosquitoes?

According the WWF, a combination of rapid warming of the globe and extreme weather events is intensifying disease outbreaks and making previously unaffected regions susceptible. Malaria and dengue fever are starting to affect new populations as warmer conditions allow mosquitoes to survive over a wider area and at higher latitudes than before.[3] The WHO estimates that about 150,000 people die each year from a combination of malaria, cholera and diarrhoeal diseases.

According to one study, if temperatures rose by 2°C (3.6°F) above pre-industrial levels, the number of people living in Africa at risk of malaria, in climatic terms, would be increased by about 40–60,000,000. Currently some 4.5 million Africans are exposed to the disease.[4]

Dengue fever brought about by climate change would expose a further 5–6,000,000,000 people, mainly in the developing world, if temperatures were to rise by 4°C (7.2°F) compared to about 3.5 billion people with no climate-change element.[5]

Global warming will therefore have a significant effect on vector-transmitted diseases as regions become more suitable for the vectors of the diseases mentioned (mosquitoes, rodents, ticks, etc.) to thrive.

Hot and humid conditions in the 1990s led to malaria cases being reported in various states in the USA, from California to New York and even Toronto in Canada.[6]

Most temperate regions, however, (for now), which include most of Europe, remain unsuitable for vector-transmitted diseases, but that could change as global warming continues to raise temperatures, which in turn increases the habitat where mosquitoes could thrive.

We now move on to Chapter W, to find out what effects global warming may have on the Earth's weather as temperatures increase.

Key points

- Vectors are organisms such as mosquitoes, which carry micro-organisms (pathogens) such as malaria, which can be transmitted to humans by the vector organism.
- Global warming will probably increase the prevalence of mosquitoes as the warmer conditions allow the mosquito to survive over a wider area and higher latitude than before.
- A temperature rise of 2°C (3.6°F) above pre-industrial levels could expose a further 40–60,000,000 people living in Africa to vector-transmitted malaria.
- Currently there is no vaccine available for malaria.

---

1 WHO (impact of climate extremes).
2 WHO, 2006 data.
3 WWF, <www.panda.org>.
4 Stern Review on The Economics of Climate Change, Part II.
5 Ibid.
6 WWF, <www.panda.org>.

W

WEATHER

It seems logical that one of the main consequences of a changing climate through global warming is going to be the weather. There is a big difference, however, between the weather and climate. Weather is really the atmospheric conditions existing at a particular time and place, whereas climate is really the long-term average weather in a particular area or place.

It could be said of the weather that 'London basked in three hours of sunshine yesterday, but today it is raining'. If you were talking about climate, however, it could be said, 'London usually gets five inches of rain in the month of February'.

Climate is therefore measured over a long period of time as opposed to what might be happening on a day-to-day basis as of course the weather is constantly changing.

As you may have read in the preceding chapters, global warming as a result of increased greenhouse gases appears to have warmed the Earth by about 0.74°C (1.33°F) over the last 100 years.

What effects might this have on the weather though, now and in the future?

The weather is of course an extremely complex system, and the main weather events that may arise as a result of global warming are likely to be the following:

Heatwaves
Floods
Droughts
Hurricanes

Heatwaves

While global average temperatures have increased by 0.74°C (1.33°F), the weather in Europe has increased by about 0.95°C (1.71°F) during the last 100 years or so, according to the European Environment Agency (EEA) and the WWF.[1]

The last fourteen years have seen eight of the warmest years in Europe's history. Warming is estimated to be greatest in southern countries like Spain, Portugal, Italy and Greece, but less along the Atlantic coastline.[2]

According to the Stern Review there will be more extreme heat days compared to today and fewer cold days, as the distribution of temperature shifts upwards. This means that if the average temperature increases by one standard deviation (equal to 1°C for some parts of Europe) the probability of today's 1 in 100 event, such as a severe heatwave, will increase by ten times, making it a 1 in 10 event![3]

This is difficult to believe, especially because as I write this chapter the UK is having one of the wettest Junes on record. But, wetter weather is also forecast, as we shall see from the section on floods later on!

Europe in 1995, and especially 2003, suffered major heatwaves, which killed some 35,000 people during 2003, 15,000 in Paris alone. The deaths in Paris seem to have been exacerbated by the urban heat-island effect, which kept night-time temperatures high. Lack of electricity also added to the problems in France, generally caused by insufficient water to cool electricity generating nuclear power plants.[4]

Heatwaves and higher temperatures will generally initially benefit countries in higher latitudes such as Canada, Russia and Scandinavia, with higher agricultural and lower mortality as cold-related deaths decline, with a 2–3°C (3.6–5.4°F) temperature rise. In the lower latitudes, however, water stress and heatwaves will affect southern Europe and California in the USA, where heat-related deaths may outstrip cold-related mortality. In Europe, Spain, Portugal and Italy are likely to be affected the most by increasing numbers of heatwaves and forest fires.[5]

In August 2007 terrible fires raged across Greece. While these appear to have been started deliberately, the very hot and arid conditions allowed the fires to spread rapidly across the mainland.

The world's major cities will feel the effect of higher temperatures as the urban heat-island effect in inner city areas amplifies the heat.

Floods

Flooding in Europe is likely to increase, especially in coastal regions of the UK. Ironically, as mentioned earlier, while writing this section, during the last week of June 2007, torrential rain has caused severe flooding in the north of England, which has turned roads into rivers in some towns and villages. While such flooding may have been caused by a shift in the jet stream over the British Isles, bringing with it excess rain, it is an example of what the scientific models predict will happen as the world warms.

In other parts of Europe, melting alpine snow and extreme rainfall may lead to more frequent floods in major river basins like the Danube, the Rhine and the Rhône. In general it seems that most scientific weather models predict increased rainfall at higher latitudes, i.e. the northern hemisphere, with reduced rainfall in the tropics and subtropical areas.[6]

Drought

As temperatures increase, reduced snowfall and shorter winters will start to affect the Pacific coast of the USA and California, together with farmlands of the Mississippi basin.

Australia will be particularly vulnerable as the Earth's driest continent. The east coast of Australia, which is home to about seventy per cent of the population and most of Australia's major cities, has suffered longer droughts and declining rainfall. It is speculated that the drought experienced by Australia, in April/May of 2007, is the first sign of the effects of global warming on a developed nation. The Queensland rainforest will also be under threat from drier and hotter summers, and tropical diseases such as dengue fever spread by mosquitoes could extend down to Brisbane and even Sydney if temperatures increase by 3°C (5.4°F) or more.[7] According to the Stern Review, one study predicts that the fraction of Earth's land area in moderate drought at any one time will increase from twenty-five per cent at present to fifty per cent by the 2090s, and the fraction of land in extreme drought from three per cent to thirty!

Hurricanes

'Hurricane' and 'typhoon' are names given to a strong tropical cyclone. A 'tropical cyclone' is a generic term for a low-pressure system that has a definitive cyclonic surface-wind circulation.

Before considering the effects global warming may have on these weather systems, we will look at a few hurricane facts and figures.

Depending where a cyclone occurs will determine whether it's called a hurricane, typhoon or tropical cyclone. Hurricanes form in the North Atlantic and Northeast Pacific Oceans. Typhoons form in the Northwest Pacific area to the east of 160° longitude. Cyclones form in the Southwest Pacific Ocean, and the North and Southwest Indian Ocean.

We will look in detail at the Atlantic hurricane season, which officially begins on 1st June and continues to the end of November each year. However, hurricanes do of course occur outside this time period. The seasons are different for Pacific and Indian Ocean areas, particularly the Northwest Pacific basin, where cyclones can occur all year round.

According to NOAA, hurricanes rotate in a counter-clockwise direction around a central 'eye', and a tropical storm will be classed as a hurricane only when wind speeds reach 74 mph (119 km/h) or more.[8]

Hurricanes can of course cause immense damage, especially if they hit land, where heavy rain, strong winds and especially strong waves – called the storm surge – can wreak destruction, as the unfortunate people of New Orleans found out during August 2005, when Hurricane Katrina hit with devastating consequences.

Hurricane strength is measured using the Saffir-Simpson scale, named after two engineers from the US National Hurricane Center, who developed it in 1969. The scale is used only to describe hurricanes that form in the Atlantic and Northeast Pacific basins.[9]

The scale has five intensities depending on wind speed. For example, a category one hurricane has a wind speed of 74 mph (118 km/h) and a category five hurricane 156+ mph (251 km/h). Hurricane Katrina, mentioned above, was probably the most devastating storm in US history, causing an estimated $100,000,000,000-worth of damage. Katrina was a category five storm, but dropped to a three when it hit land on the eastern seaboard of the USA, in August 2005.

Atlantic basin hurricane numbers

According to NOAA, hurricanes have been monitored since 1944 by aircraft reconnaissance, though data does exist as far back as 1899, owing to the highly populated coastline of the eastern USA, which lies in the path of hurricanes that form in the Atlantic basin. NOAA have records showing the annual hurricane figures going back as far as 1851, but data obtained since 1944, when aircraft monitoring began, is regarded as more reliable, until satellites became available in the 1970s.[10]

These statistics show that five or more *major* hurricanes (Saffir-Simpson scale three, four or five) occurred over four different years during the 1950s, three different years during the 1960s and 1990s, and over two years so far during the 2000s, with 2004 seeing six major hurricanes and 2005 seven. The 2006 season was quieter, with two major hurricanes reported.

The 2007 NOAA outlook called for a likely range of thirteen to sixteen named storms, seven to nine hurricanes, and three to five major hurricanes. It is believed that this actually translated to five hurricanes with two being major, the same as in 2006.

Year 2005, however, was the most active hurricane season since reliable records began, with a record twenty-eight named storms, fifteen of which were hurricanes, with seven of them major. This beat the previous record held in 1969 with twelve hurricanes, five of which were major.

Year 1950 was also good for Atlantic hurricanes, with eleven, eight of which were major!

What do the statistics show?

Well, the period from 1995 to present shows a definite increase in hurricane activity, in both intensity and frequency, in the Atlantic basin. The figures show that the 1950s, 1960s and the 1990s were very active hurricane periods. From 1995 to 1999, however, there have been a record thirty-three hurricanes, but the four years prior to that, 1991 to 1994, saw the quietist period since the 1940s. Scientists have found that there appears to be a cycle for mid-Atlantic hurricanes that alternates between quiet and active periods over a twenty- to twenty-five-year period. It is possible therefore that the heightened activity since 1995 is just another active period rather than anything to do with global warming.

Typhoons in the Northwest Pacific basin, it seems, have also increased in intensity since about 1980, but that was after a similar decrease in activity between 1960 and 1980, a similar activity cycle to that of Atlantic hurricanes.[11]

So, while there appears to be a multi-decadal cycle with hurricanes and tropical storms, it is difficult at this point to pin the increase in frequency and intensity since 1995 on global warming.

What about sea temperature?

According to NOAA, it is likely that the quiet decades of the 1970s to the early 1990s, regarding major Atlantic hurricanes, were due to changes in Atlantic Ocean sea-surface temperature structure, with cooler than usual waters in the North Atlantic. The reverse situation of a warm North Atlantic was present during the active late 1920s through to the 1960s.[12]

Hurricanes are fuelled by warmer water, which helps them grow. Therefore it would be logical that as the Earth warms, causing increased sea temperatures, stronger hurricanes may emerge.

NOAA's position however is that it is highly unlikely that global warming has (or will) contribute to a drastic change in the number or intensity of hurricanes.

On the other hand global modelling studies suggest the potential for relatively small changes in tropical cyclone intensities related to global warming. NOAA also mentions, in its 2005 Atlantic hurricane season summary, that characteristics of an active multi-decadal signal in the Atlantic include

> 'Warmer surface sea temperatures…'

> 'An African easterly jet (AEJ) that is favourable for primarily the development and intensity of tropical disturbances moving westward off the coast of Africa…'

and interestingly,

> 'Recent studies indicate that in addition to the multi-decadal oscillation, the destructive power of hurricanes has generally increased since the mid 1970s when the period of rapid increase in global ocean and land temperatures began.'[13]

It is also interesting that 2005, according to NASA, was the warmest year on record globally, which coincided with a record year for North Atlantic hurricanes, with twenty-seven named storms, fifteen of which were hurricanes, and seven of which were major, with an unprecedented four reaching category-five status. Coincidence? Maybe.

Emily, Katrina, Rita, and Wilma

Emily formed in July and was the first category-five storm of the 2005 season. Emily was also the earliest category-five storm on record. Emily was just one of five named storms for July, again the most *named* storms on record for the month of July. NOAA do not have a record of how much damage Emily caused when she hit Mexico on about 18th July 2005.

Then we have Katrina. There aren't many people who haven't heard of Hurricane Katrina. Katrina developed in August 2005 and is likely to be one of the most devastating and costly storms in US history. Katrina made landfall in Louisiana and Mississippi at category-three strength, having reduced from a category-five storm. Katrina caused 1,300 deaths and possibly as much as $125,000,000,000 in damage, making it the most costly hurricane in history.

Then we have Rita, which developed in September 2005. Hurricane Rita came close to the Florida Keys on 20th September, causing storm-force winds of up to 76 mph (122.3 km/h) on Key West. Rita then strengthened and tracked towards the Gulf of Mexico and reached category-five status on the Saffir-Simpson scale, with wind speeds of up to 175 mph (281.6 km/h) and proceeded to cause an estimated $10,000,000,000 in losses.

Finally Wilma arrived in October 2005, the last category-five storm of the 2005 season. Wilma produced about 60 inches (152.4 centimetres) of rain while passing over the Yucatun Peninsula in Mexico, and made landfall in Florida as a category-three storm, causing an estimated $12,000,000,000 in losses![14]

Luckily the 2006 season was much quieter. August 2007 saw the first category-five storm of the season. This was Hurricane Dean, which caused significant damage to Jamaica as it made its way towards Mexico, before reducing in strength.

How do hurricanes form?

As mentioned earlier, for hurricanes to develop, certain environmental conditions must be present, such as warm ocean water, high humidity and favourable atmospheric and upward spiralling wind patterns off the ocean surface. Atlantic hurricanes usually start off as a weak tropical disturbance off the West African coast, and intensify into rotating storms with weak winds called tropical depressions. It's only when wind speeds reach at least 74 mph (118 km/h) that they are classified as hurricanes. NASA scientists and NOAA have been studying how winds and dust conditions from Africa influence the birth of hurricanes in the Atlantic Ocean, using an armoury of NASA's Earth observing satellites.[15]

Scientific proof of severe weather link to global warming?

It seems that NASA is on course to establish whether or not global warming is indeed influencing the weather. NASA's AIRS instrument on board the Aqua satellite, which is short for Atmospheric Infrared Sounder, can measure very subtle changes in the Earth's climate. Scientists from NASA and NOAA, as well as other scientists, have already demonstrated that AIRS data can lead to better forecasts about the location and intensity of 'extratropical cyclones', which are mid-latitude storms, often striking the east coast of the USA. This may enable the team to test the climate-weather hypothesis once more data is available. AIRS will also have the ability to test the hypothesis that climate change may be causing the water (hydrological) cycle to accelerate, by measuring the humidity distribution within the atmosphere. This will show with sufficient accuracy whether the water cycle is indeed speeding up. If so, as is suspected in a warmer world, there will be more water vapour and clouds in the atmosphere resulting in more rainfall. If so, AIRS will be able to establish a link between global warming and the weather, as a faster water cycle causes greater rainfall as a result of an accelerated hydrological cycle.[16]

El Niño and La Niña

Not a set of Spanish twins, but El Niño and La Niña are names given to the opposite phases of an oscillation of the ocean atmosphere system in the tropical Pacific Ocean, which can affect weather around the world. El Niño, which most people have heard of from weather reports, is characterised by significantly warmer-than-normal waters in the Pacific Ocean, which can extend all along the equator. El Niño events seem to occur roughly every three to five years, and are usually followed (but not always) by La Niña events, which produce the opposite effect, water temperatures colder than normal. Typically it seems that El Niño events occur more often than La Niña events.[17]

During an El Niño event, temperatures in the winter are warmer than normal in the northern central USA, and cooler than usual in the southeast and southwest. During a La Niña year, winter temperatures are warmer than normal in the southeast and cooler than normal in the northwest.

The name El Niño did actually derive from Spanish fishermen off the coast of South America, who noticed unusually warm waters in the Pacific Ocean at the start of the year. El Niño means 'the little boy' or 'Christ child' in Spanish. La Niña means 'the little girl' and is the name given to a cold event, the opposite of El Niño.

The terms ENSO (El Niño Southern Oscillation) and ENSO cycle are used to describe the full range of variability observed in the southern oscillation index, including both El Niño and La Niña events.

Impacts of extreme weather events

We have looked at how much damage hurricanes can cause. However, according to the Stern Review on The Economics of Climate Change, the consequences of climate change are going to be felt earliest and most strongly through changes in extreme events, such as storms, floods, droughts and heatwaves. Indeed, the floods of June and July in the UK are already estimated to cost in the region of £2,000,000,000, and this figure is sure to rise.

Annual losses are put at about $60,000,000,000 since the 1990s, and record costs of $200,000,000,000 were racked up in 2005, the year Hurricane Katrina hit the USA. Analysis from the insurance industry has shown that weather-related catastrophe losses have increased by two per cent each year since the 1970s over and above wealth, inflation and population growth/movement.[18]

In the next chapter we will look at what will happen to Earth's species as the consequences of global warming take effect, in Chapter X, Xtinction.

Key points

- ➢ The last fourteen years have seen eight of the warmest in Europe's history, and in 2003 alone some 35,000 people died from heatwaves.
- ➢ Flooding in Europe is likely to increase, especially in coastal regions of the UK. In April/May 2007 Australia experienced a drought, which may be the first signs of global warming and its effects on a developed nation.
- ➢ 'Hurricane' and 'typhoon' are names given to a strong tropical cyclone. They are identified by reference to part of the world in which they form.
- ➢ The Atlantic hurricane season starts on 1st June and officially lasts to the end of November.
- ➢ Hurricane strength is measured using the Saffir-Simpson scale, which has five intensities – the first denotes a hurricane with wind speeds reaching seventy-four mph.
- ➢ Year 2005 was the most active hurricane season since reliable records began, with fifteen hurricanes, seven of which were major.
- ➢ In the Atlantic basin, hurricane activity in both intensity and frequency has increased since 1995. NASA's AIRS instrument on board the Aqua satellite will have the ability to establish whether there is a link between global warming and the weather, by looking to see if the water cycle has speeded up.

---

1 WWF (climate change and extreme weather events in Europe, summer 2005 article).

2 Ibid.

3 Stern Review on The Economics of Climate Change, Part I.

4 Op cit, Part II.

5 Op cit (key messages).

6 Op cit (summary of regional impacts of climate change).
7 Ibid.
8 National Oceanic and Atmospheric Association, <http://hurricanes.noaa.gov>.
9 National Oceanic and Atmospheric Association, <http://www.aoml.noaa.gov/hrd/tcfaqHED.html>.
10 Ibid.
11 Ibid.
12 Ibid.
13 Ibid (climate of 2005 hurricane season).
14 Ibid.
15 NASA, <www.jpl.nasa.gov> (July 2006 news release).
16 NASA, <www.airs.jpl.nasa.gov>.
17 NOAA, <www.panel.noaa.gov> ('What is an El Niño?').
18 Stern Review on The Economics of Climate Change, Part II (impacts of extreme events).

Hurricane Emily

Hurricane Emily had come ashore in Mexico on 20th July 2005, when the Moderate Resolution Imaging Spectroradiometer (MODIS) on NASA's Aqua satellite captured this image.

Credit NASA, <www.nasa.gov>.

Hurricane Rita

Credit NASA, <www.nasa.gov>.

Hurricane Katrina
Credit NOAA, <www.katrina.noaa.gov>.

Hurricane Wilma
Credit NOAA, <www.noaanews.noaa.gov>.

X

XTINCTION

This chapter title should of course be spelt with an E, but it didn't seem appropriate to deal with this subject matter so early in the book! The most severe consequences of global warming will result in species extinction, and once a species becomes extinct, that is the end of it. If animals and plant species are becoming extinct as a result of global warming, could humankind be next?

Biologists are now making a link between global warming and the threat to Earth's biodiversity as species become extinct.

Two papers published in 2004 in the journal *Nature* by Chris Thomas and his colleagues analysed the distribution of 1,103 species of animals and plants from various parts of the world, and the results showed that fifteen to thirty-seven per cent are likely to become extinct based on the best projections of future climate change to 2050.

Dr Thomas points out, however, that not all species committed to extinction will disappear by 2050, due to the inherent time lag of extinction.

> 'When the climate becomes unsuitable for the long term survival of species, it does not mean that it will die off immediately. For species with long lived individuals, in particular it may take many decades or even centuries before the last individuals die out, so, these are the numbers of species that may be declining toward extinction from 2050 onwards, but not the numbers that would have died out by that date.'[1]

According to the WWF, this report has now been expanded on in a study published in the scientific journal *Conservation Biology*, which finds that global warming represents one of the most pervasive threats to our planet's biodiversity. This study picks up where the *Nature* study papers left off, incorporating critiques and suggestions from other scientists while increasing the global scope of the research to include diverse hotspots around the world.

The results reinforce the massive species extinction risks identified in the 2004 study. Its lead author, Dr Jay Malcolm, said,

> 'Climate change is rapidly becoming the most serious threat to the planet's biodiversity. This study provides even stronger scientific evidence that global warming will result in catastrophic species loss across the planet.'[2]

Lessons from history

Earth has been through at least five mass-extinction events. Starting with the oldest first, there was the Ordovician extinction that occurred about 439,000,000 years ago. Then we have the Devonian extinction, about 354,000,000 years ago, and the better known extinctions of the Permian, the Triassic, the Cretaceous and last but not least, the Palaeocene-Eocene Thermal Maximum (PETM), extinction event. These events occurred 252,000,000, 200,000,000, 65,000,000 and 55,000,000 years ago respectively. Most people are aware of the event 65,000,000 years ago which wiped out the dinosaurs when an asteroid collided with Earth. The PETM event was looked at in detail in chapter H, so starting with the Permian we will look at the most recent extinction events in more detail, simply because more is known about them.[3]

Permian Triassic event (P-T event)

About 252,000,000 years ago, when the Earth had only one continent, called Pangea, all multi-celled life on Earth was nearly wiped out. This event represents the greatest dying-out in the fossil record, when more than ninety per cent of species were lost. The NASA Astrobiology Institute reported in 2005 that results from South Africa provided the best ever picture of the P-T extinction on land, suggesting that it was a much more complex process than would be expected from a comet or asteroid impact, which is almost certainly how the dinosaurs reached their demise some 197,000,000 years later.

Peter Ward and his colleagues, from the University of Washington in Seattle, presented the first detailed study of vertebrae extinction patterns from the later Permian and early Triassic periods. They found that various species of therapsids, 'mammal-like reptiles', became extinct at different times, consistent with extended climate or habitat change. The contending Permian extinction theories of induced climate change posit an increase in temperature (perhaps caused by an enhanced atmospheric greenhouse effect) and lower levels of oxygen. This may have been the result of large-scale volcanic eruptions from what is now Siberia. It may be no coincidence that huge Siberian eruptions took place at about the same time. A factor that cannot be ruled out by scientists, it seems, is that an asteroid or comet impact caused the extinction, though no impact craters have yet been found that would explain this event.[4] According to NASA, however, a possible submarine impact structure has been located off Western Australia.

It is believed that temperatures may have increased by 10–30°C (18–54°F), higher than they are today.

Whatever the cause, the Permian-Triassic event signalled the end for most species, including the top land predator of the time, the dimetradon, an 11-foot (3.3-metre) long, 250-pound reptile, which had a fin-like sail on its back.

Triassic-Jurassic event (T-J event)

About 52,000,000 years later, as the dawn of the dinosaurs approached (about 200,000,000 years ago), another mass extinction occurred. This one is called the T-J event, as it happened on the boundary of the Triassic-Jurassic periods and killed off about fifty per cent of all of Earth's species. It seems the last of the mammal-like reptiles were killed off, but the early dinosaurs were spared. Evidence for this was again found by the palaeontologist Peter Ward, of the University of Washington, from the fossil record of the collapse of single-celled organisms called protists. The extinction is thought to have happened in fewer than 10,000 years, a very short time in geological terms.

Although the scientific paper did not speculate on what caused the event, an asteroid or meteor is considered likely due to the speed (in geological terms) at which species became extinct.[5]

Cretaceous-Tertiary event (K-T event)

Most people are aware of this extinction event, as it is the one that wiped out the dinosaurs some 135,000,000 years later, 65,000,000 years ago. The event is the only one that has left behind a definitive smoking gun in the form of a crater. This can be found in the Yucatan Peninsula situated in the Gulf of Mexico, caused by an asteroid thought to have been a few miles in diameter impacting on the Earth. The dinosaurs were wiped out, but this extinction event more than likely paved the way for modern *Homo sapiens* to evolve from their earlier ancestors, some 64.8 million years later![6]

So it seems mass extinction of species on our planet, probably brought about by asteroid or comet impacts, has occurred on average every 93.5 million years. It is now 65,000,000 years since the last extinction event, which is a little worrying when considering it was only 52,000,000 years between the P-T and T-J events, but this does not necessarily mean we are overdue for the next asteroid or comet impact!

To find out what NASA is doing to try and prevent another asteroid/comet collision extinction event from occurring, please see the end of this chapter.

So, it is clear that the Earth has gone through five major extinction events in the past. Could global warming bring about the sixth extinction event?

It seems that most biologists agree that we are already in the midst of a sixth great extinction event as a result of climate change and global warming.[7]

Climate change can affect species in a number of different ways, for example, through increased incidence of disease, change of habitat, temperature and precipitation changes and loss of food availability.

According to the WWF, animals and plants more suited to cooler climates will have to move from their usual habitats poleward, a process already being observed in the alps, mountainous regions of Queensland in Australia, and in the forests of Costa Rica.

Fish stocks have also been observed moving from their usual habitats in the seas around Cornwall off the British Isles, northward to the Shetland and Orkney Isles.

Animals under threat

The polar bear

The consequences for all Arctic species will be devastating as the Arctic starts to melt. The polar pear (*Thalarctos maritimus*) uses the ice to hunt for seals and fish. However, as the sea ice is now melting earlier in spring and forming later in the autumn, the time the bears have on the ice is diminishing. In the southernmost range of the bear's habitat, in the Hudson and James Bays of Canada, scientists have found that the main cause of death in polar bear cubs is either a lack of food or the lack of fat on nursing mother bears. As the periods without food lengthen, due to the melting ice, the overall body condition of the polar bear declines, which means it has less stored fat to use as energy. This then affects the bear's ability to last through the warm season when there is less ice and little available food.

Polar bears are having a hard time. There seems to be some dispute going on as to whether the polar bear is under threat from climate change at all. This wasn't helped by a picture of polar bears looking distraught on a melting iceberg, which was used by the press to highlight their plight. It turned out the photo was taken during the summer, when of course the sea ice melts in any event. This revelation has had the effect of seeming to reduce the threat global warming has on Arctic polar bears. Nevertheless, it is estimated that about 20–25,000 bears are left in the wild, sixty per cent or so living in Canada. With Arctic ice melting so rapidly, it's difficult to see how their general survival for the aforementioned reasons is not under threat. A useful website from the WWF for keeping up to date with polar bears in relation to climate change can be found at <www.panda.org/polarbears>.

Sea turtles

Marine sea turtles are also being affected as sea levels rise, leading to erosion of their nesting beaches. Warmer ocean temperatures often lead to coral bleaching and other damage to coral reefs, which are essential turtle-feeding habitats.

Whales

The North Atlantic right whale, which resides in the Atlantic off the coast of the USA, is also under threat as warming waters contain less plankton, which means less food for the whales to feed on. The lack of availability of food is becoming an increasing cause of mortality. Only about 300–350 of these whales exist, with little hope of population growth.

Giant panda

Chinese panda bears are also being threatened as their forest habitat becomes fragmented. The panda's staple diet, bamboo, is also part of a delicate ecosystem that could be affected by the changes caused by global warming.

Orang-utans

These are Asia's only apes, and they are in big trouble! It is believed that this animal could become extinct within a few decades, as its habitat, the rainforests of Indonesia, becomes threatened by deforestation and climate change. Global warming increases the duration and frequency of droughts, putting even more pressure on the ape's already heavily logged forest home.

Elephants

African elephant habitats are under threat as humans take more and more of their land away, thus giving elephants little chance to escape any changes to their natural habitat caused by global warming.

Tigers

There are only about 6,000 tigers left on Earth. Poachers, lack of prey and loss of habitat are all factors adding to the tiger's demise. The largest remaining areas in which tigers reside are the mangrove forests of India. Projected sea-level rises could wipe out these habitats altogether.[8]

Birds

Global warming appears to be having a significant effect on the bird species. A recent report compiled for the WWF reviewed more than 200 scientific articles. It seems there is a clear and escalating pattern of climate-change-impact on bird species around the world, suggesting a trend towards major extinction from global warming. Particularly vulnerable birds are migratory, mountain, island, wetland, Arctic, Antarctic, and seabirds.

Bird species that can move easily to new habitats will not necessarily be affected, however.

Scientists have found decline of up to ninety per cent in some bird populations, together with total and unprecedented failure in others.

If global warming exceeds the 2°C (3.6°F) threshold above pre-industrial levels, bird extinction rates could be as high as thirty-eight per cent in Europe and seventy-two per cent in Northeast Australia.[9]

Pikas

Pikas are small hamster-sized flower-gathering mammals, and relatives of rabbits, that live in the alpine regions of the western USA, southwestern Canada, and Asia. They may be one of the first mammals in North America to fall victim to global warming.

They usually collect and then dry their foods in the summer sun in distinctive 'hay piles' ready for winter use, as food is difficult to obtain during the winter in the alpine environment.

Pikas live in the rocky areas of these alpine regions, and they are particularly vulnerable to temperature change. According to the WWF, research indicates global warming may have contributed to local extinctions of pika populations, which may be the 'canary in the coal mine' when it comes to the response of alpine and mountain systems to global warming.[10]

Frogs

Australian frogs are being hit hard as a result of their freshwater homes drying out from drought, possibly caused by global warming. Some frogs also need water to be able to breed and therefore any reduction in suitable breeding grounds or rainfall will impact on this species.

Another problem has been caused by a certain fungus that appears to have wiped out a species of toad. The golden toad from the forests of Costa Rica has been killed off by this fungus, which has thrived as the conditions in the forest have become more suitable, brought about by climate change.

Penguins

Just as the icons of the Arctic, polar bears, are feeling the heat from global warming, so are the penguins of the Antarctic. According to the WWF report on Antarctic penguins and climate change, the four populations of penguin that breed on the Antarctic continent – Adélie, emperor, chinstrap and gentoo – are under escalating pressure. For some, global warming is taking away precious ground on which penguins raise their young. For others, food has become increasingly scarce because of warming in conjunction with over-fishing. The Antarctic peninsula is warming five times faster than the average rate of global warming, and the vast Southern Ocean has warmed all the way down to a depth of 3,000 metres.

Coral reefs

Anyone who has ever been scuba diving or snorkelling will know how beautiful coral reefs are, together with all the marine life that makes the reefs their home and feeding grounds.

Coral reefs are home to some twenty-five per cent of all marine life, and are the most biologically diverse ecosystems within the ocean, rivalled only by the Earth's tropical rainforests.

The corals are also marine animals and the coral itself lives in a symbiotic relationship with an algae called zooxanthellae, thriving in the coral animal tissue, and carrying out photosynthesis providing energy for themselves and the coral. It is the algae that give the coral its amazing colours.

When environmental conditions become too stressful the zooxanthellae algae may leave their coral homes, depleting the corals of energy and colour, a process referred to as coral bleaching.

Coral bleaching may arise as a result of the following climatic change impacts:

1. Warmer sea temperatures. Coral bleaching can be caused if sea temperature increases by as little as 1–2°C (1.8–3.6°F) above average annual maximum temperatures.

2. Acidification of the world's oceans causes decreases in the coral's ability to grow and respond to other stresses.

3. Stronger storms can cause havoc to coral reefs.

Once the coral has become bleached, the only way for it to survive is if it is recolonised by the zooxanthellae in a reasonable period of time.[11]

While all these animals are or will be affected by global warming, it's probably the image of the polar bear swimming across Arctic waters devoid of sea ice in search of food and solid ice that is now the most recognisable.

Only time will tell whether it's too late to save these animals now facing possible extinction from a warming world.

If we can't protect these animals from extinction, then what chance does humankind have? It would be arrogant to assume that we would not suffer the same fate at some point in the future if nothing is done.

Spaceguard

While global warming is a threat to species over time, should an asteroid or comet somewhere out there with Earth's name on it impact the planet, then the human species, together with almost everything else, could be wiped out in an instant. It has become apparent over the last decade or so that near-Earth objects (NEOs) are a significant threat to all life on Earth, and the Spaceguard program is in place to try and locate and if necessary protect the Earth from such a threat.

In 2005 NASA passed the Authorization Act, also known as the George E Brown Jr Near Earth Object Survey Act. The objectives of the programme are to detect, track, catalogue and characterise the physical characteristics of NEOs equal to or larger than 140 metres in diameter, with a perihelion distance of less than 1.3 au (astronomical units) from the sun. Another objective is to achieve a ninety per cent completion of the survey within fifteen years of enactment of the NASA Authorization Act of 2005.[12]

If you recall Chapter M, perihelion refers to the closest approach of an object as it orbits the sun. One astronomical unit represents in general terms one Earth-sun distance, taken as 150,000,000 kilometres or 93,000,000 miles.

A range of options is considered for diverting such a threat, including nuclear standoff explosions, slow-push techniques, and non-nuclear kinetic impactors. Nuclear explosions run the risk of fracturing the object into smaller parts, all of which could continue on their paths to Earth. Slow-push techniques would be costly, and mission times of many years, if not decades, would be needed to divert an incoming object from colliding with the Earth, if it was on a collision trajectory. Non-nuclear kinetic impactors are seen as the maturest approach, and could be used in some deflection/mitigation scenarios, especially for NEOs that consist of a single small, solid body.

The worrying part to all this is that currently the NASA Act carries out the Spaceguard survey only to find NEOs greater than one kilometre in diameter, and this programme is currently budgeted at about $4.1 million (about £2 million) per year for years 2006 to 2012.

NASA continues to gather knowledge and gain experience in its ability to tackle potential threatening asteroids, by virtue of its Discovery space series of missions, such as Near-Earth Asteroid Rendezvous (NEAR), Deep Impact and Stardust missions.

NASA launched its Dawn mission on 27th September 2007, to investigate the two largest known main belt asteroids, Vesta and Ceres, which lie between the planets Mars and Jupiter. Dawn will commence exploration of these asteroids in 2011 and 2015 respectively.[13]

At the present time NASA recommends that the programme be continued as currently planned. However, due to budget constraints, NASA is unable to initiate any new programmes at this time.

Therefore at present no one on Earth is searching for asteroids with a size of less than one kilometre, with a potential impact trajectory with Earth, which is a very worrying thought!

Spaceguard UK

The British government also set up a task force on potentially hazardous NEOs, which was established in 1999.[14] A report published in 2000 made fourteen recommendations, which can be found on the internet:

<www.nearearthobjects.co.uk/report/pdf/
full_report.pdf>

It took a four-year campaign by Spaceguard UK, and a concurrent political drive by Lembit Opik MP, before the British government announced the establishment of a task force on NEOs. The members are Dr H H (Harry) Atkinson (chairman), Professor David Williams and Sir Crispin Tickell. Their terms of reference were to confirm the nature of the impact hazard, identify current UK activities, and make recommendations on future action.[15]

While it seems the UK is well aware of what needs to be done, and is encouraging other nations and organisations to take an interest, there appears little prospect at this time of a coordinated approach to the problem, and the USA is the only country on Earth with an active programme in mitigating the NEO hazard.

While asteroid impact is a threat that could wipe out all life on Earth, it is something that countries have the resources to tackle if the desire is there. While the cost of such programmes is great, it is insignificant compared to the expense of the current war in Iraq, or even the 2012 Olympic games!

Just as global warming is now finally being recognised as a major threat to all life on Earth, it may take an errant asteroid coming slightly too close to Earth for comfort before significant funding is channelled into NEO search programmes!

Investment now to mitigate the problems of global warming and to search for NEOs will be as nothing to the future cost when the consequences of such threats materialise.

After this fairly depressing chapter, in the next we will look at what little changes we can all make as individuals to reduce our carbon footprints and what governments are doing to try and save planet Earth from global warming and its consequences – Chapter Y, You Can Help.

Key points

- ➢ Earth has been through at least five mass extinction events in the past.
- ➢ Biologists are of the view that we are currently in the midst of a sixth great extinction event as a result of climate change and global warming.
- ➢ In 2005, NASA put in place Spaceguard, a programme designed to track and detect rogue asteroids that pose a threat to planet Earth.

1 Mongabay, <www.mongabay.com>.
2 WWF, <www.panda.org>.
3 Space.com, <www.space.com> (on the five worst extinctions in Earth's history, Lee Siegal).
4 NASA Astrobiology Institute, <www.nci.arc.nasa.gov>.
5 Ibid (on asteroid and comet impact hazards).
6 Ibid (on asteroid and comet impact hazards).
7 Mongabay, <www.mongabay.com> (lessons from historical extinctions).
8 WWF, <www.panda.org> (on threatened species, polar bears, tigers).
9 WWF, <www.panda.org> (on climate-change impact on bird species).
10 WWF, <www.panda.org> (on threatened species, pikas), and <www.worldwildlifefund.org>.
11 WWF, <www.worldwildlifefund.org> (on corals and coral science).
12 NASA, <www.neo.jpl.nasa.gov> (near-Earth object program).
13 NASA, <www.down.jpl.nasa.gov>.
14 <www.spaceguarduk.com>.
15 NASA, <neo.jpl.nasa.gov>.

Opposite above: The Arctic polar bear will become increasingly threatened as its habitat becomes affected by global warming as more and more ice starts to melt as Arctic temperatures increase. Many other animals are threatened with extinction as a result of global warming.

Opposite below: Spaceguard was set up by NASA in 2005 to protect Earth from a rogue asteroid which, if it collided with Earth, could cause a global extinction event.

Y

YOU CAN HELP

We are almost at the end of this alphabetic journey on global warming. We have looked in detail at the causes of global warming, the evidence, the consequences and relevant related issues. We will now look at what we can all do as individuals to reduce the amount of CO_2 in the atmosphere in an effort to save planet Earth from overheating, and perhaps give scientists and governments enough time to avert the predicted worst-case scenarios if levels of CO_2 and other gases keep rising unchecked.

This chapter will also help you calculate your own carbon footprint, and will also explain carbon offsetting and look at various UK code-of-conduct-compliant companies that now enable you to offset the carbon you produce, for a small fee of course! Some American companies will also be looked at.

You may be thinking that there isn't much that you can do as an individual to tackle the seemingly enormous problem of global warming. On the basis that the UK is responsible for just under two per cent (550,000,000 tonnes) of CO_2 annually, it may seem like a pointless exercise, as realistically the UK's contribution is minimal. On the other hand if everyone does her bit to help, then vast amounts of CO_2 and other greenhouse gases can be kept out of Earth's atmosphere.

While the UK is responsible for just under two per cent of world CO_2, the EU's twenty-nine countries are collectively responsible for about 14.5 per cent, and the USA alone a massive 19.8 per cent. If momentum can be achieved, it's possible that two per cent can progress to ten per cent, then twenty, and so on. So, if everyone in every nation does his or her bit to help, maybe, just maybe, we can make a difference.

This chapter will tell you how to do it, because YOU CAN HELP!

What is a carbon footprint?

Just in case you're wondering, it is not the mark left on your carpet after you have stepped into a pile of coal or ash! We are all responsible for the release of CO_2 into Earth's atmosphere. Some of us are guiltier than others.

Almost everything we do, whether as individuals, large companies or governments of countries, causes heat-trapping CO_2 and other greenhouse gases to be released into the atmosphere.

Take a typical day as an example

We wake in the morning, perhaps turn on the radio or television, usually having left it in standby mode from the night before. Most of us then take a bath or shower, put the kettle on, turn on some lights, maybe make some toast. Finally before going to work we perhaps put the dishwasher on or maybe the washing machine. This is all before the day has even started! Most of us will then get into our cars and drive to work. When in work our computers go on, the lights go on and more tea and coffee gets made! When we get home in the evening, the televisions go back on, computers, games consoles, mobile phone chargers, before finally the lights get put out before we go to bed, until the next day.

If you wander about your house at two a.m., how many red or green standby lights would you find illuminated?

Each and every appliance in the house, whether it is actually on or just left on standby, uses electricity. If it's using electricity then more than likely a coal-fired electricity-generating power station somewhere not too far away is belching out CO_2 into the atmosphere.

Every time we travel anywhere by car, train, boat or plane we add CO_2 and other greenhouse gases into the atmosphere.

The amount of CO_2 we cause to be emitted into the atmosphere as a result of our actions and lifestyle over a set period of time, say one year, is our carbon footprint.

So, if you want to take action to help Earth's environment, this is what you can do to help.

It wouldn't be practical or much fun to stop going on holiday suddenly, or stop watching television or drinking coffee for that matter, but the following tips will save money and more importantly reduce your individual carbon footprint at the same time.

At home

Avoid standby

Turn off those appliances when they are not in use – phone chargers, TVs, and DVDs are the main culprits. These appliances can use anywhere between ten and sixty per cent of their normal power when on standby. Make sure the plugs of these appliances are easily accessible so that they can be turned off easily. It takes only a few seconds to do it!

It is estimated that about 3,000,000 tonnes of CO_2 is emitted as a result of standby wastage annually in the UK.

Better still, invest in the following:

Standby-buster and energy tracker

You can now buy these two very nifty little gadgets that can both save you money and CO_2 emissions. The standby-buster looks just like a timer switch, but it's basically a gadget you plug in, and then plug the socket of your TV or DVD or any electrical appliance into it. This then allows you to turn off the appliance at the wall remotely. You can use the remote handset to turn off all electrical appliances in the house if you want, provided each appliance is plugged into a standby-buster. The equipment functions via radio waves, so there's no problem using it through walls or between floors! This eliminates having to reach for those tricky plugs to turn them off at the wall. The gadget claims to have a payback time of six months.

The gadget can be purchased at <www.allthingsgreen.net> and <www.amazon.co.uk>.

The energy tracker measures how much energy your electronic equipment is consuming, both when the equipment is on and in standby mode. Therefore you can see which electrical items are consuming most power. The only way to ensure that no power is being consumed, however, is to turn the appliance off when not in use.

The gadget can be purchased at hardware stores such as B&Q.

Appliances

Try and buy energy-efficient appliances. Avoid placing your fridge next to the cooker as it will have to work harder to keep things cool! It is obvious, but don't turn on the dishwasher, washing machine or tumble dryer unless it is full. Buy the most energy-efficient appliances when replacing items.

If based in Europe check the EU energy label that appears by law on appliances, and the website <www.topten.info> may help.

If based in the USA, check for the energy star label and energy guide labels.

Lighting

Turn off the lights when you are not in the room. Replace your bulbs with energy-saving fluorescent bulbs. In the older incandescent light bulbs most of the power goes to producing heating rather than light. Low-energy light bulbs use a third of the power and give off the same amount of light. While they may be a little more expensive they will last about four times longer!

Boiling the kettle

Boil enough water for one or two cups, if making a drink only for yourself. Electricity and CO_2 emissions will be saved if the kettle isn't trying to boil when full.

Turn down the heating

By turning the household thermostat down 1°C (1.8°F) you will save money, reduce your carbon footprint, and you probably won't even notice the slight drop in temperature in any event!

Use the sun

If you can afford to, have solar panels put on the roof of your house to provide your hot water, and photovoltaic panels to provide electricity.

Insulate properly

Make sure the loft space and water tank are insulated properly to prevent unnecessary heat loss.

Take a shower

Taking a shower uses a lot less energy than a bath. If you are used to having a bath every day, why not try and cut down to once or twice a week?

Recycling

Recycling rubbish plays a very important role in reducing energy requirements needed for creating new products from scratch and disposing of waste that could be used again. Here are some simple recycling facts:

* Of 35.5 million tonnes of waste in the UK in 2003/4, only seventeen per cent was recycled – source, <www.defra.gov.uk>. Some European countries recycle up to fifty per cent of their waste.
* Seventy per cent less energy is needed to make recycled paper as opposed to making it from raw materials.
* 12.5 million tonnes of paper and cardboard are used each year in the UK.

- ✻ Twenty-four trees are needed to make one tonne of paper.
- ✻ One recycled tin can saves enough energy to power a television for three hours.
- ✻ Sixty per cent of rubbish in a dustbin could be recycled.
- ✻ Unreleased energy from the average dustbin each year could power a television for 5,000 hours (2.7 years if television used five hours a day)!
- ✻ Aluminium cans, if recycled, can be ready to use in about six weeks.
- ✻ Glass bottles and jars that are thrown away end up on landfill sites and never decompose, but glass is 100 per cent recyclable and can be used again and again and again.
- ✻ Plastic can take up to 500 years to decompose and can easily be recycled.

A very useful UK website, which tells you how to recycle, where your local recycling sites are, and who runs your recycling services can be found at <www.recyclenow.com>.[1]

Go that extra mile

Why not get all your energy from renewable energy sources? If you live in Europe check out the following website to find out your nearest green energy supplier:

<www.eugenestandard.org>.[2]

In the UK the WWF have teamed up with Ecotricity, which is a green energy supplier. Ecotricity allows you to sign up for either 'new energy' or 'new energy plus'. The first option means that a portion of your electricity (about twenty-six per cent) comes from renewable sources, in fact from Ecotricity's own wind turbines. The rest comes from conventional means, but every pound you spend will go towards building new wind turbines to increase the amount of electricity over time from renewable sources.

Ecotricity will match the cost of your usual local supplier on this tariff.

The second option, 'new energy plus', is where the electricity provided is 100 per cent green. Electricity not generated from Ecotricity's own wind turbines comes from existing clean hydro, solar and wind sources. This tariff will be about five per cent more expensive as a result.

Why not switch? The 'new energy' option won't cost you any more and you know your money is going towards investing in future renewable energy projects.

Check out their website at <www.ecotricity.co.uk>.[3]

Did you know?

When purchasing new electrical appliances, a new European WEEE directive (waste, electrical and electronic equipment) is in force (as of, in the UK, 1st July 2007), which will help you recycle your old electrical and electronic equipment. The legislation covers the whole of the EU, and its purpose is to reduce the amount of electronic waste at landfill sites.

Manufacturers, retailers and importers of electronic goods are obliged to put systems in place that allow customers to recycle their unwanted electrical goods free of charge. The directorate covers virtually every piece of electrical equipment you can think of, from televisions and computers to freezers and electronic toys. Just look for the wheelie-bin sign with a cross on it on the packaging of electrical items and ask your local store if they will take your old goods in for recycling.

Travelling

Automobiles are the worst offenders when it comes to CO_2 emissions within the transport sector. Public transport is of course less polluting per head, so getting the bus, train or walking will reduce significantly your carbon footprint. Changing your vehicle for a hybrid, or better still all-electric, when readily available, would be a great way of reducing transport-sector emissions. Aeroplanes, while currently lagging behind emissions from automobiles, contribute to a rapidly increasing sector as airports all over the world expand to cater for growing numbers of passengers. Instead of jetting off to Europe or elsewhere, why not take the train? Any journey with a flight time of say up to three hours can easily be matched in convenience and speed when taking the train directly into the heart of most major cities, and your carbon footprint will be much reduced. Only recently Ryan Air have been hauled up according to advertising standards as they suggested that it was quicker to fly to Brussels from London than it was to get the Eurostar. This was simply untrue as the Ryan Air plane lands somewhere outside the city, so by the time you get off the plane and get a train, bus or taxi into town it is probably quicker to get the train in the first place!

How about a carbon friendly vacation?

Try booking a holiday with Responsible Travel.[4]

While some people don't agree with carbon offsetting on the basis that you can pay to alleviate your guilt as a polluter, it seems logical to be able to offset the CO_2 we produce *once* efforts have been made to cut down or reduce the amount of CO_2 we are directly responsible for as individuals, as we go about our daily lives. Clearly carbon offsetting should be considered as a supplement to reducing one's carbon footprint and not something to do in addition to producing whatever carbon emissions you want!

The above tips show us how to take some small steps to reduce our carbon footprint as we go about our daily lives.

Have you ever wondered what level of *greenhouse gas emissions* we are actually responsible for annually?

- The average UK citizen produces about eleven tonnes.[**]
- The average US citizen produces about 24.3 tonnes.[**]
- The average global citizen produces about 4.4 tonnes.[*]
- There are about 6.4 billion tonnes for the US as a whole.[**]
- There are about 655,000,000 tonnes for the UK as a whole.[**]

* There are about 43 billion tonnes for the planet as a whole.**

* The Climate Trust offset figures.[5]
** CAIT year 2000 figures.[6]

Carbon offsetting

We will now look at how to go about offsetting the CO_2 that we can't reduce or prevent. There are various companies that allow you to calculate CO_2 emissions as a corporation or an individual, whether from your home, individual trips (car, plane, train or boat), or just your entire annual CO_2 emissions. That then enables you to make a payment to offset those emissions. These companies will usually either invest your money in a renewable energy project somewhere, such as reforestation projects to sequester (capture) CO_2 from the atmosphere, or a wind farm, or some other worthy renewable project, thus offsetting or counteracting an amount of CO_2 that would otherwise have been emitted into the atmosphere, with the idea of reducing the effect of your own emissions to zero. So offsetting basically

Reduces amounts of CO_2 in the atmosphere.
Provides funds for renewable energy technologies.
Reduces the effect of your emissions to zero.

We will look at an assortment of companies and the projects that they invest in and also compare CO_2 emission offsets to look at how the cost varies.

Gold standard

The WWF and other organisations developed the gold standard, setting the marker for truly additional projects that also benefit sustainable development. It is basically a set of quality-control criteria for carbon-offset projects. As discussed in Chapter K, the Kyoto mechanism allows for carbon offset projects called the Clean Development Mechanism (CDM) and the Joint Implementation projects (JI). The purpose of the gold standard is to ensure these projects are both reducing CO_2 emissions and fostering sustainable development. It is the view of the WWF that the gold standard reflects best the objectives of the CDM as defined in the Kyoto Protocol.

Gold standard projects should give confidence to host countries and the public that projects represent new and additional investments in sustainable energy services.

The gold standard is actually a Swiss-based non-profit organisation. Credits can be purchased in the retailer list of the marketplace section. However, the Pure Trust company mentioned below is just one in a number of companies supporting the gold standard.

A list of compliant organisations can be found on the gold standard website, <www.cdmgoldstandard.org>.

Virgin Atlantic for example now allow you to purchase carbon offsets from their website when booking a flight, should you wish to do so, through an organisation called MyClimate, a company supporting the gold standard criteria.

In the UK a new code of practice was due to be launched in the autumn of 2007, setting a quality mark for offsetting companies to comply with, which will ensure that the companies are meeting a certain standard when it comes to offsetting carbon emissions. There are some companies that already comply with the scheme:

Pure, the Clean Planet Trust.
Global Cool.
Equiclimate and Carbon Offsets.

An American company, called Clean Air-Cool Planet, also prepared a report that identified a number of providers, both in the UK and USA, performing best against the reports methodology. The following is a list of some of those providers:

Carbon Neutral Company (UK).
Climate Care (UK).
CO2balance (UK).
Climate Trust (USA).
NativeEnergy (USA).

We will have a brief look at each of these providers and see what they have to offer.

PURE, THE CLEAN PLANET TRUST

This company's website can be found at <www.puretrust.org.uk>. The company is a registered charity, with the specific objective of improving air quality to combat climate change. Offsetting schemes are monitored by independent trustees, and donations to Pure benefit Kyoto Protocol emissions reductions projects around the world. One of their current projects is the Malavalli power plant project in India, which will reduce CO_2 emissions by about 20,000 tonnes a year, and supplies about 10,000 people with clean, renewable electricity in about forty-seven villages.

Emission reductions are audited and verified by the United Nations Clean Development Mechanism, or in the UK the National Energy Foundation.

Carbon credits purchased that could otherwise be used by polluters within the Kyoto emissions trading schemes are cancelled from the register or retired, which then limits the amount of greenhouse gases polluting companies are allowed to release into the atmosphere.

Pure, according to their website, is the UK's leading carbon offset scheme.

GLOBAL COOL

This company's website can be found at <www.globalcool.org>.

Global Cool's mission is to get 1,000,000,000 people to reduce their personal CO_2 emissions by at least one tonne. A donation to prevent a tonne of CO_2 from getting into the atmosphere will cost you just £20 ($40).

There is even a breakdown on the website showing how your donation gets spent: £14 of it goes directly to high-quality energy-reduction projects (which spares your tonne of CO_2 from Earth's atmosphere) and alternative energy technology companies to speed up the uptake of solar-, wind- and biomass-generated power.

A counter on the company website counts down from

1,000,000,000 tonnes of CO_2 entering the atmosphere as donations are made.

Equiclimate

This company's website can be found at <www.ebico.co.uk>.

Equiclimate is a UK government-approved carbon offsetting service. You can complete a few simple questions, which will indicate how much CO_2 you are personally responsible for. Your own Kyoto target is calculated, telling you by how much you need to reduce your emissions. Equiclimate, through EBICo, then enter the European emissions trading scheme and purchase CO_2 allowances on your behalf, equal to the amount of CO_2 you are responsible for. These are then retired and kept out of the market rather than sold to other traders and brokers, to ensure that the UK's cap on emissions is reduced by the amount of the allowance purchased.

CarbonNeutral Company

This company's website can be found at <www.carbonneutral.com>.

The CarbonNeutral Company has been in the carbon offsetting business for some time, and it was the first company in the world to create an online carbon calculator to make it easy to measure CO_2 emissions. The company's core service is the sourcing and supply of carbon-offset projects from worldwide technology and forestry projects.

As with the other companies, you can use an online carbon calculator, whether you are an individual or a business, and your money will go towards emission reduction projects and forestry projects that sequester (absorb) CO_2 already in the atmosphere. The company works primarily with the following types of projects:

Renewable energy.
Energy efficiency.
Methane capture.
Forestry.

By purchasing a carbon credit, for every tonne of CO_2 that you are unable to reduce, you can pay for its reduction somewhere else in the world.

This company also lets you choose which projects you want to invest in. For example, *to offset emissions through travelling on a return flight to New York from London, you will pay between £9.00 and £12.90. This is based on your flight covering 11,082 kilometres and producing 1.2 tonnes of CO_2*. The projects you can invest in are as follows:

1 Future portfolio package – £12.90 (invests in new technologies while saving the CO_2-equivalent of the flight).
2 International communities portfolio – £9.88 (this helps with projects in developing countries and will also save the same amount of CO_2 as the flight produces).
3 One World portfolio – £9.00 (combination of world technology package and UK forestry, which will again save the same amount of CO_2 as your flight produces).

You also get a certificate, map and information about the projects invested in, and even a recycled leather bag!

This company also has a team of people to measure, reduce and set a strategy to meet agreed commercial objectives. If you are a large company and don't know where to start making cutbacks on CO_2 emissions or offsetting, a team of advisors is on hand to help.

CLIMATECARE

This company's website can be found at <www.climatecare.org>.

Climatecare was set up in 1998 and is a trading name for Climate Care Trust Ltd. The trust does not have any shareholders and is committed to leading a best-practice approach to offsetting emissions. Some arguments have been raised in general about offset companies planting trees to sequester CO_2 from the atmosphere. While this alone isn't the answer to increasing levels of greenhouse gases, it is estimated that about twenty per cent of CO_2 emissions annually come from deforestation and forest fires. Climatecare therefore aim to have twenty per cent of their CO_2 liabilities in reforestation to balance these emissions. The current project is in the Kibale National Park in Uganda, and lots more information about this can be found on Climatecare's website.

After royalty costs of ten per cent are paid to Climatecare's start-up company, ninety per cent of any monies made go to Climatecare's projects and staff and office costs. Money paid gets put into a pool to fund a portfolio of projects in order to reduce CO_2.

Climatecare's calculators also work out the amount of greenhouse gases in general an activity such as flying produces as opposed to just CO_2. As we know from Chapter G, greenhouse gases are converted to CO_2 equivalents and referred to as CO_2e. Emissions from certain activities, such as flying, will have a greater effect at altitude, and Climatecare is using the latest science to take this into account.

Climatecare will calculate CO_2e amount and express the figure as tonnes of CO_2.

The cost of offsetting one tonne of CO_2 with Climatecare is currently £7.50.

Climatecare's offset calculator allows you to calculate anything from a holiday, driving, even to a wedding attended by 150 guests!

Using the same example above, to offset emissions from a return trip to New York from London would cost you £11.55. Resulting emissions from the trip are calculated at 1.54 tonnes of CO_2 for travelling 6,911.8 miles.

Once you have made an offset, Climatecare will provide a certificate confirming that. You can also buy gift sets for family and friends if you like.

Climatecare is also monitored by eminent environmentalists, including the WWF, to ensure that offset projects achieve the CO_2 emissions claimed, so any offsets purchased meet the strictest of criteria.

CO2BALANCE

This company's website can be found at<www.CO2balance.com>.

Founded in 2003, this company invests in both projects that

save energy emissions and absorb CO_2, usually by way of forestry-based offsets. All offsets offered by this company are in projects controlled by it. For example, CO2balance will plant offset trees only on their own land, to ensure that the long-term future of the woodland is secured, and will harvest the trees before they die to ensure that the carbon absorbed by the trees during their lifetime is not released back into the atmosphere. The wood from the trees is then used in construction projects to ensure the carbon is locked up in the wood, beyond the life of the tree itself.

This company offers offsets only for new projects rather than renewable energy projects, to avoid selling carbon rights as opposed to actually funding the development of such projects.

CO2balance operates according to the concept of 'additionality', meaning that the majority of offset funds come from carbon-offsetting clients, as the company feels it is useless to invest in a project that would have happened anyway, when the client has made no difference. For this reason they will never sell you carbon rights in an existing project.

As with the other companies, you can choose to offset your holidays, household or personal CO_2 emissions, and the offset calculator also allows you to make a choice as to which project you can invest in, from replacing light bulbs in Kenya to a reforestation project in a Cornish village in the UK.

Using our example above, the cost of offsetting the CO_2 from a return flight from London Heathrow to New York's JFK Airport would be £23, for an offset of 2.55 tonnes of CO_2.

CO2balance uses a radiative forcing of two for all aviation offsets, to take into account the environmental effects of aircraft emissions at altitude, which are greater than the effects of burning fossil fuels alone, as previously discussed.

THE CLIMATE TRUST

This company's website can be found at <www.climatetrust.org>.

The Climate Trust is an offset company based in the USA, and during its time it has offset more than 2.7 million metric tons of CO_2 from $9,000,000 invested in offset projects. It is one of the largest and most experienced offset buyers in the USA.

The company's offset criteria are:

1. An offset project needs to be one that would not otherwise occur without the funding provided by the offset purchaser.
2. Results must be rigorously quantified.

The trust offers offset projects in many sectors, which include green buildings, industrial efficiency, forest sequestration, transportation, and of course renewable energy.

All the projects can be viewed on the company's website, and you can also obtain a certificate confirming your offset purchase.

*Native*Energy

This company's website can be found at <www.nativeenergy.com>.

Another US company, *Native*Energy offers a choice of offsetting between traditional renewable energy credits and offsets from operating new projects. The company is Native American majority-owned, and purchase and sale records are reviewed by an independent certified public accountant.

As with the other companies, *Native*Energy allows you, via their carbon calculator, to purchase offsets from activities ranging from driving various-sized cars to heating, electricity, or simply offset however many tons of CO_2 you like.

Detailed information on all their projects can also be found on their website.

Should you wish to offset CO_2 and greenhouse gases generated as a result of your lifestyle, you can, through companies like these, as individuals and corporations alike, both in the UK and the USA (and of course other countries have a great deal of choice and variety).

Rainforests

As mentioned in Chapter A, anyone who wishes to make a donation, either to save an acre of rainforest, or for a carbon sequestration project, and which involves reforestation and rainforest protection in various locations around the world, such as Ecuador, Belize and India, can do so through the World Land Trust.

Their website is <www.worldlandtrust.org>.

The minimum amount of CO_2 that can be offset is 0.33 tonnes, and this would cost you £5, or about $10. Certificates are also issued on request for donations of £15 (about $30) or more.

The website calculator also allows you to offset CO_2 emissions from your everyday life as well, from travel to household emissions.

WWF policy

The WWF accepts the latest scientific evidence that steep reductions in greenhouse gases need to be made by 2050, in the order of sixty to eighty per cent, to avert dangerous warming over and above 2°C.

Its US policy efforts include the following:

1. Building support for a new global agreement on including forest-based emissions in the post-2012 global climate agreement. The WWF will focus on measures that reduce deforestation and degradation, while maintaining the potential to reforest degraded habitats as well.
2. Working to cap US emissions. Working with institutions to advance legislation that commits the USA to effective reductions in emissions.

Large companies lead the way

Leading companies have also joined forces with the WWF to reduce their emissions voluntarily by joining WWF climate savers. Climate savers companies will reduce their collective emissions by about 10,000,000 tons of CO_2 each year by 2010. This is equivalent to taking 2,000,000 cars off the road. Companies participating in this scheme are as follows:

Johnson and Johnson
IBM
Xanterra parks and resorts
Polaroid
Nike
Catalyst
Lafarge
Tran-Sport Communication SAGAWA
Novo Nordisk
Tetra Pak
Sony

For more information on these companies, and how they have achieved their reductions in CO_2, you can take a look at WWF US website at:

<www.worldwildlife.org/climate/projects/climatesavers/companies.cfm>

In the UK, Sainsbury's, Asda and Bodyshop also recently announced that they would stop selling products containing non-sustainable palm oils, to prevent tropical rainforest deforestation. Such a measure could help prevent orangutans from becoming extinct. Almost ninety per cent of palm oil comes from Malaysian and Indonesian rainforests, home to the ape.

Marks and Spencer have also recently announced a five-year programme with the aim of becoming carbon-neutral by that time.

HRH Prince of Wales

Prince Charles is also leading by example, and is well known for his environmentally friendly living, having launched his own organic food products from his farm, near Highgrove, and conscious and open about his impact on the environment when it comes to climate change. The Prince has also introduced a number of climate-friendly measures at his farm, including solar panels for heating water. He was recently awarded the Global Environment Citizen Award for his environment achievements.

Past Global Environmental Award-winners include Edward O Wilson, Harrison Ford, Jane Goodall, Bill Moyers, and Al Gore.

What are governments doing?

Apart from the Kyoto Protocol and the EU's emissions trading scheme, which has already been discussed in Chapter K, we will look at the UK, USA and the rest of the world briefly in turn.

UNITED KINGDOM

The UK government is aiming to reduce its own travel and general CO_2 emissions by questioning the need for its ministers and civil servants to travel in the first place, if video and telephone conference facilities could be used instead.

The government's carbon offsetting scheme (GCOF) will then be used to calculate CO_2 emissions that cannot be avoided, as a result of official air travel. Emissions will be calculated, and then multiplied by two to take into account the radiative forcing caused by aircraft emissions at altitude, as discussed in Chapter G.

To offset these emissions, it is government policy to purchase emission reduction credits (ERCs), which are carbon credits from CDMs. These, if you remember, were discussed in Chapter K, and are procedures under the Kyoto Protocol by which developed countries can finance projects in developing countries that reduce CO_2 emissions and as a result receive CO_2 credits for doing so.

For more information on the UK government's emissions reduction policy visit <www.defra.gov.uk>.

Energy White Paper

The UK's international and domestic energy strategy is set out in the Energy White Paper, and one of its main goals is to cut CO_2 emissions by some sixty per cent by about 2050, with real progress towards this goal by 2020. One of the ways this will be achieved is a mandatory emissions trading scheme, referred to as a carbon reduction commitment (CRC). The scheme will apply to organisations, such as hotel chains, supermarkets and universities etc., which consume over 6,000 megawatt hours of electricity a year. That's 6,000 megawatts or 6,000,000 kilowatts. Remember, an average house over one year would consume about 8,500 kilowatt hours, while 6,000 megawatt hours of electricity is equivalent to an electricity bill of about £500,000 a year in the UK. The system would work as a cap-and-trade scheme, with the government setting a global amount of maximum energy for the organisations within the scheme, which would allow the organisations to receive a bonus if they used less than their allowance and be given a penalty if they exceeded it.

The White Paper also discusses the role nuclear energy will play in the UK's low-carbon economy. More information can be found at <www.dtistats.net>.

Renewable transport fuels obligation

This is a measure to ensure that by 2010 at least five per cent of all fuel sold at UK garage forecourts is from renewable sources. This policy will ensure that about 1,000,000 tonnes of CO_2 is kept out of the atmosphere.[7]

Climate change bill

As mentioned in Chapter K, the UK has taken the lead by introducing a bill, which if passed will make the UK the first country in the world to set legally binding targets to reduce its CO_2 emissions. The target, using many methods as discussed above, will be a sixty per cent reduction by 2050. The bill should receive royal assent in the autumn of 2008. A draft copy of the bill can be found at the following website:

> <http://www.officialdocuments.gov.uk/document/cm70/7040/7040.pdf>

WHAT ABOUT THE USA?

While the USA is *the* world's major polluter (possibly just overtaken by China) and a non-signatory to the Kyoto Protocol, the country is taking steps to reduce its emissions and is investing millions of dollars into researching new science and technology to reduce greenhouse gas emissions. No doubt, just as in Australia, once President Bush leaves power, the country will also sign up to the protocol, but only time will tell.

Energy Policy Act 2005 (EPA)

The above Act became law on 8th August 2005. It seeks to provide a long-term strategy for the US's energy challenges in an environmentally friendly way. President Bush also announced two initiatives to run alongside the EPA, which are the Advanced Energy Initiative (AEI) and the American Competitiveness Initiative (ACI). The former proposes to increase investment significantly in alternative fuels and clean-energy technologies, and the latter recognises the need for funding research programmes in physical sciences over the next decade.

The result of the EPA and the above initiatives is as follows:

1 Development of new biorefineries
Advance the development of such refineries for the production of biofuels, bioproducts, biomass heat and power. Twenty-seven new ethanol plants have come online since the enactment of the EPA, and US ethanol production is expected to increase to about 8,000,000,000 gallons by end 2007.

2 Clean coal projects
Clean coal project funding for the Clean Coal Power Initiative (CCPI) will focus on accelerated coal research and new technologies to help use coal, which is the cheapest and most abundant fossil fuel, as discussed in Chapter F. Carbon sequestration technology is researched and this will be looked at further a little later in this section.

3 Geothermal power
There has also been a surge in geothermal power projects in the USA, with some forty-five projects now under development in nine different states. When developed these projects could deliver between 1778 and 2055

megawatts of new electric power. Geothermal energy was looked at if you recall in Chapter R.

4 Solar power
Also proposed was the Solar America Initiative (SAI). This initiative will accelerate the development of advanced solar electric technologies, including photovoltaics, and concentrating solar power systems with the aim of making them cost-competitive with other forms of renewable energy by 2015. It is expected that about 1,000,000 homes will be powered from solar energy, which will reduce CO_2 emissions by about 10,000,000 metric tons per year.

5 Nuclear power
Nuclear power is the only mature technology with potential to supply large amounts of power without emissions of CO_2 and other pollutants. The USA has not licensed a new nuclear power plant for over thirty years! The EPA however recognises that more nuclear power plants are required in the USA and around the world. Research is being carried out into the next generation nuclear plant (NGNP), which will provide improved technologies over existing plants, and co-generation of hydrogen by nuclear energy. Generation IV International Forum (GIF) is a multilateral partnership of ten countries and the European Commission that is fostering international cooperation in research and development for the next generation of nuclear power plants. These will be designed to produce electricity, hydrogen and other energy products with much less waste and zero air pollutants and greenhouse gases.[8]

6 Fusion energy
Energy from fusion is considered the holy grail of energy production. Fusion, as opposed to fission, is the reaction that occurs within our sun and the stars. The process relies on fusing atoms, rather than splitting

> them, which is the process in conventional nuclear reactors. A major benefit over other forms of electricity generation is that hydrogen can be produced without carbon emissions.

The US Department of Energy is involved in the fusion energy science programme, which will allow investigation into the physics of burning plasma at high densities in an experiment known as Iter (Latin for 'the way'). The purpose of the experiment is to demonstrate that fusion can be used to generate electrical power with a view to collecting the necessary data to design and operate the first electricity-producing plant.

Such a project involves huge challenges, due the extremely high temperatures needed to create nuclear fusion.

Work is ongoing with six other partners – China, the EU, India, Japan, Russia and South Korea – to build a fusion reactor based in Cadarache, France.[9]

FutureGen

The FutureGen project is another innovative quest, which will create the world's first zero-emissions fossil-fuel plant. This would be a major breakthrough, bearing in mind energy production from fossil fuels is going to be with us for decades to come. Energy from burning coal, as we know from Chapter F, is the biggest single source of CO_2 emissions. Roughly thirty-three per cent of the USA's and about forty-five per cent of the world's carbon emissions come from coal-fired power plants and other large-point sources. Carbon sequestration, carbon capture, separation and storage or reuse, will play a major role in reducing CO_2 emissions from burning fossil fuels such as coal.

FutureGen will cost about $1.5 billion to build, but when operational the prototype plant will be the cleanest fossil-fuel-fired power plant in the world, producing both electricity and hydrogen from coal, while capturing and sequestering the CO_2 generated in the process. As the plant also produces hydrogen, it is hoped that it will lead to the creation of a hydrogen economy and fuel pollution-free vehicles.

The project will require about ten years to complete, but the results will be shared among all participating countries and the industry as a whole.[10]

There is concern, however, that it simply may not be practical to store all the sequestered CO_2 once it has been captured. Once captured the CO_2 is converted to liquid form, which then must be safely stored somewhere!

Global Warming Solutions Act

California is setting the standard in the USA for reducing emissions by passing Assembly Bill 32, or as it is known the Global Warming Solutions Act. This was performed in 2006 by California's Democrat-led assembly, with Republican Arnold Schwarzenegger's backing. This legislation creates an obligation on California to cut emissions to 1990 levels by 2020, using a cap-and-trade system similar to the EU emissions trading scheme. California amazingly is the world's twelfth largest source of CO_2 emissions![11]

It seems that other states are following suit, and it is expected that a greenhouse gas emission cap-and-trade system may be introduced nationwide.

CHINA

China has also released a report pledging its aim to reduce energy use by twenty per cent by 2010. It intends doing this by increasing the use of nuclear, wind and other forms of renewable energy, though not at the expense of sustaining development and eradicating poverty, which is seen as the country's first priority. China is prepared to enter into agreement to cap its carbon emissions when the Kyoto successor agreement comes into force, provided the USA is also included, which seems entirely reasonable! Will the USA continue to put up obstacles, however?

REST OF THE WORLD

United Nations

UN Secretary Ban Ki-moon convened a high-level event that took place in New York on 24th September 2007. Its purpose was to promote discussions on ways to move the international community towards negotiations on a new global agreement on climate change, at the UN climate-change conference, which took place in Bali on 3rd December 2007.

The purpose of discussions was to try and get in place a multilateral framework for action on climate change for the period after the Kyoto Agreement ends in 2012.

Ban Ki-moon has stated that he intends to take a leadership role on the issue of climate change, and also help the international community address the problem. Three special envoys have been appointed to assist with consultations with governments on how best to progress these negotiations for a successor to Kyoto.[12]

These discussions simply must succeed for there to be any chance of curbing Earth's CO_2/greenhouse gas levels.

Asia Pacific Economic Cooperation (APEC)

A September 2007 meeting in Sydney, Australia, of APEC-member countries, which include Russia, China, Australia, Japan and the USA, managed to agree on two matters: to reduce energy intensity (amount of energy needed to produce a dollar of gross domestic product) by twenty-five per cent by 2030, and to increase forest-cover in the region by at least 50,000,000 acres (200,000 square kilometres) by 2020.

While the pledges made have been criticised because the goals are voluntary and no targets set for reducing greenhouse gases, the agreement is significant owing to the participation of the USA and Australia, which were both non-signatories to the Kyoto Agreement (Australia ratified 3rd December 2007), and China, which is largely exempt from Kyoto on the basis that it is still a developing nation. The USA, China, Russia and Japan are also the largest emitters of greenhouse gases globally.

Wedge strategy for preventing global warming

Two Princeton professors, Rob Socolow and Stephen Pacala, introduced a concept, which was first published in the 2004 edition of *Science Magazine*. The concept is fairly simple. In order to stabilise emissions over the next fifty years, the world needs to reduce emissions of carbon by about seven gigatons (7,000,000,000 tons). Fifteen stabilisation wedges were identified, with each wedge reducing emissions by one gigaton each.

So, the idea is to break down the seemingly insurmountable goal to reduce emissions into fifteen smaller, more easily achievable wedges. Examples of each wedge are as follows:

1 Increasing fuel economy in cars from thirty mpg to sixty mpg, and cutting down the distances cars travel by fifty per cent – two wedges.
2 Cutting carbon emissions in buildings and appliances by twenty-five per cent – one wedge.
3 Increasing the number of one-megawatt wind turbines by 100 times current capacity – two wedges.
4 Increasing current US or Brazilian land available for ethanol production by 100 times, equivalent to a sixth of Earth's cropland – one wedge.
5 Introducing carbon-capture and sequestration on 800 gigawatts-worth of coal plants (about 800 plants), and improving the efficiency of coal-fired plants – three and one wedge respectively.
6 A 700-fold expansion of solar-generated electricity – one wedge.
7 Preventing deforestation and increasing reforestation – one wedge.
8 Vastly increasing electricity production from nuclear and gas power plants – two wedges.
9 Conservative ploughing of cropland. As farmers turn over soil, CO_2 escapes into the atmosphere. If this practice could be increased huge amounts of CO_2 could be prevented from escaping from tilled soil – one wedge.

There we have the fifteen wedges. While some of these methods appear fairly straightforward, each wedge would take a considerable amount of global effort to achieve. Having said that, generation of power from wind turbines is growing at about thirty per cent annually. For its wedge to be achieved by 2054 it needs to grow by only about eight per cent annually, so already, achieving this wedge doesn't seem such a hurdle as long as growth continues in this sector, which it appears to be doing.

Only seven wedges need be achieved, not all fifteen, which also makes the goal seem a little easier.[13] As mentioned above, however, it will take the USA and other participating countries at least ten years to build its FutureGen project, which will no doubt create and assist the technology required to make serious inroads into the coal-plant efficiency and carbon-capture and sequestration wedges.

Plenty of things are being done, and still can be done, to prevent global warming from destroying life on planet Earth as we know it. While the task appears insurmountable, surely we must at least try to prevent the worst predictions that global warming warns us with. Are you prepared to make a difference?

In the next chapter we will consider how time is not on our side. The clock is ticking for planet Earth and its inhabitants. It is zero hour.

Key points

- The UK is responsible for just under two per cent of Earth's warming greenhouse gases, the USA almost seventeen per cent.
- Your carbon footprint is a term used for the amount of CO_2 you are personally responsible for emitting into Earth's atmosphere over a fixed period of time.
- Almost everything we do requires electricity, most of which comes from coal-fired electricity power stations.
- Carbon offsetting is a way of counteracting any emissions you produce, effectively reducing your emissions to zero, if done properly.

- ➢ The USA is developing FutureGen, which will be the world's first zero emissions fossil fuel power plant.
- ➢ The wedge strategy is a plan to tackle global warming by assigning fifteen identifiable stabilisation wedges to a variety of environmentally friendly methods of reducing CO_2 emissions in order to reduce global emissions by at least 7,000,000,000 tons over the next fifty years. The idea is that this seemingly insurmountable task can be made more manageable by focusing on each wedge separately.
- ➢ Nuclear energy will play a significant role in helping to provide Earth's future electricity demands as it produces zero greenhouse gas emissions.

---

1 <www.recycle.guide.org.uk>.
2 <www.eugenestandard.org>.
3 <www.ecotricity.co.uk>.
4 <www.responsibletravel.com>.
5 <www.climatetrust.org>.
6 <www.cait.wri.org>.
7 Road Transport Fuels Obligation, <www.dft.gov.uk>.
8 US Department of Energy, <www.energy.gov>.
9 Ibid.
10 Ibid (FutureGen).
11 A32 Global Warming Solutions Act 2006.
12 United Nations, <www.un.org>.
13 World Resources Institute, <www.wri.org>.

Z

ZERO HOUR

Finally we reach Chapter Z. I say 'finally' as this is the end of a twelve-month-long journey of research and investigation into all the issues surrounding global warming and the damage that human beings are causing the environment and atmosphere of planet Earth. Having been uncertain about the facts and confused about the causes and effects of global warming before, I am now certain that warming of Earth's delicate atmosphere is taking place and on course to cause unpredictable and possibly catastrophic damage to the planet and all life upon it.

The purpose of this final chapter is to remind us of the facts and evidence discussed in the preceding chapters, and to confirm that global warming is real and that humans are responsible for it. This chapter will emphasise the fact that time has really run out for the Earth, and action needs to be taken now to avoid the worst consequences of global warming.

The facts

Supposing Sir Arthur Conan Doyle's famous detective, Sherlock Holmes, was investigating the case for global warming, and was asked by his trusty sidekick Dr Watson how he concluded that global warming was real and was more than likely caused by human activities. His response would be 'Elementary, my dear Watson!'

Here's the evidence:

1 Levels of Earth's naturally occurring greenhouse gases since about the 1750s have been increasing, in particular CO_2, which has increased by twenty-two per cent since accurate measurements began in 1958, and thirty-seven per cent in total since 1750.

2 The consequences of the Industrial Revolution meant that for the first time in history humankind started producing anthropogenic greenhouse gases, mainly CO_2, which coincides with an increase in CO_2 levels, particularly since the start of the twentieth century. During the latter half of the twentieth century fossil-fuelled electricity-generating power plants and automobiles and other forms of transport have been belching out CO_2 and other gases continuously.

3 We know levels of greenhouse gases, particularly CO_2, have been increasing since the mid-eighteenth century, as evidence from ice cores, and more recently direct measurements from Mauna Loa observatory in Hawaii, have scientifically confirmed this fact.

4 The Earth's rainforests, which act as natural carbon sinks, have decreased in size from twelve per cent to five per cent of Earth's land surface and continue to be destroyed by logging, deforestation for agricultural and building use, and more recently for planting crops for biofuel production. The rainforests store carbon, and as they are cut down, some of the oldest trees, which may be hundreds of years old, release their stored carbon back into the atmosphere. As the rainforests disappear, they are no longer able to capture CO_2 from the atmosphere, or indeed produce oxygen.

5 Worldwide average temperatures have increased by 0.74°C (1.33°F) during the last century, confirmed by average air, land and sea measurements. Temperature increases are more pronounced at Earth's polar regions. World sea levels have risen by 3.6 centimetres (1.41

inches), confirmed by NASA satellites. All over the planet, ice caps, glaciers and ice shelves are melting, confirmed by both NASA and the National Snow and Ice Date Centre (NSIDC). In fact this year has shown a record loss of Arctic sea ice at the end of the 2007 summer period, beating the previous all-time low of September 2005. At the end of the 2007 summer, compared to the record loss in 2005, Arctic sea ice has decreased again by an area the equivalent of five times the size of the UK.

6 The sun's role in global warming, while clearly a complex one, does not appear to be responsible for the current warming trend we are seeing. While the sun has had a major effect on climate (the Little Ice Age/ Maunder Minimum), it would seem that measurements by NASA satellites of the sun's total solar irradiance (TSI) of 0.05 per cent per decade cannot alone account for the current warming trend. This is supported by calculations contained in the IPCC reports, which suggest that the sun has a radiative forcing of 0.12 watts per square metre, compared to the radiative forcing measured from anthropogenic causes of 1.6 watts per square metre.

I don't think you have to be as cunning as Sherlock Holmes to come to the conclusion that global warming is REAL!

Paradigm energy shift needed

As mentioned in Chapter F, four fifths of CO_2 emissions that go towards heating up planet Earth come from burning fossil fuels. To avert disaster and prevent planet Earth from over-heating, together with its inhabitants, a paradigm shift in energy production needs to take place.

The only way this can be achieved at the present time is by switching energy production to renewable energy sources such as nuclear, solar, wind, geothermal and biomass as discussed in Chapter R.

More and more renewable energy projects are indeed being developed, but if the massive amounts of funding available were channelled into renewable energy sources the Earth may be saved from becoming more like its uninhabitable neighbour, Venus!

The world will spend an estimated $16 trillion on energy infrastructure between 2001 and 2030, and this is money that could be used in the transition from fossil-based fuels.

The USA also currently channels about $150,000,000,000 of its annual subsidies into fossil-fuel energy production, and this needs to be redirected to pay for cleaner energy systems.[1]

Dilemma ahead?

We know that the Earth's population is forecast to increase by about forty per cent between now and the year 2050.

Energy demand is predicted to increase by about fifty to sixty per cent by the year 2030.

Easily accessible oil is running out. According to peak oil theory put forward by the geophysicist M King Hubbert (1903–1989), oil production has already peaked and is now in terminal decline. He predicted in 1956 that US oil-production would peak between 1965 and 1970. It peaked in 1970. While the price of oil may skyrocket as supplies decline, the environment may benefit if it forces us to develop cheaper, cleaner, renewable forms of energy production.[2]

The problem, however, is that coal supplies are abundant. If oil supplies dwindle, then the Earth's enormous coal reserves may be used to fill the gap. There is enough coal left to last about 200 years or so, enough to increase Earth's temperatures to extinction levels.

There appear to be no easy answers. We can only hope that new technology being developed for the FutureGen project, discussed in Chapter Y, or possibly the development of a viable fusion reactor, will help provide the people of Earth with clean or cleaner energy. Perhaps some scientist somewhere will come up with a remarkable invention to solve Earth's ever-increasing CO_2/greenhouse gas problem before it's too late.

Time is running out for planet Earth, as it's widely accepted there is a window of about only ten years to make significant cuts in greenhouse gas emissions to avoid severe climatic change.

Realistically I suspect that only a multiple-fronted attack on global warming will help solve the crisis. In other words, only if attempts are made to reduce greenhouse gas emissions using all the methods mentioned, and maybe even additional help from an as yet unknown invention/non-polluting power source, will we succeed in preventing catastrophic global warming from making large parts of the Earth uninhabitable.

If only scientists could come up with something like an artificial tree which could mimic the action of real trees, or something similar? Maybe in the not too distant future large artificial forests will be built on the outskirts of cities consisting of super-efficient photosynthetic cells capable of replicating natural photosynthesis?

Virgin Earth challenge – hope on the horizon?

On the 9th February 2007 Sir Richard Branson and Al Gore launched the Virgin Earth Challenge, a $25,000,000 global science and technology prize to anyone who can come up with a commercially viable design resulting in a net removal of anthropogenic greenhouse gases from Earth's atmosphere. The design would have to remove greenhouse gases year after year over a ten-year period, without harmful effects, and contribute materially to the stability of Earth's climate. More on Richard Branson's pioneering challenge can be found on the Virgin Earth website.[3]

We can only hope the prize will be claimed in the not too distant future!

Potential consequences of global temperature increase above pre-industrial levels

1°C rise – Atlantic thermohaline circulation starts to weaken.

2°C rise – fifteen to forty per cent of species facing

extinction according to one estimate/potential for Greenland ice sheet to melt irreversibly.
3°C rise – twenty to fifty per cent of species facing extinction according to one estimate/some models predict onset of Amazon forest collapse/rising risk of collapse of West Antarctic ice sheet.
4°C rise – potentially thirty to fifty per cent decrease in water availability in southern Africa and the Mediterranean/7–300,000,000 more people affected by coastal flooding each year/fifty per cent of Earth's nature reserves cannot fulfil objectives.
5°C rise – disappearance of large Himalayan glaciers, affecting twenty-five per cent of China's population and hundreds of millions in India/sea-level rise threatens major cities such as London, New York and Tokyo.
Above 5°C rise – this level of global-temperature rise would be the equivalent to the amount of warming that took place between the last Ice Age and today, making it difficult to predict the effects as such temperatures would be far outside human experience.[4]

In the above scenarios 1°C represents a temperature range of between 0.5 and 1.5 degrees and so on.

Conclusion

Well, if you have managed to get this far, it is hoped you have learnt a considerable amount about all the issues surrounding global warming. Since putting pen to paper, twelve months ago, global warming and climate-change-related stories continue to appear in the press at ever-increasing rates, it seems.

The purpose of researching and writing this book was to educate myself, and to provide, in as-easy-to-understand a format as is possible for this topic, to as many other people as possible, the evidence, facts and consequences of global warming.

If by reading this book you decide to make small changes that will result in less CO_2 being emitted into the atmosphere, whether it be by reducing your carbon footprint, offsetting CO_2 emissions or ideally both, then at least some small steps will have been taken to avert the deepening crisis that is slowly enveloping planet Earth.

For my part, I have of course changed my behaviour since researching and writing this book.

I now ensure all electrical appliances not in use are switched off at the plug socket, instead of left on standby. I have purchased the standby-buster mentioned in Chapter Y to help with this.

I try and travel by train whenever possible, certainly when going to Europe.

I offset the CO_2 I produce annually, whether it be from flying, driving or household energy use.

I recycle my rubbish.

I will donate ten per cent of the net proceeds of sale of this book, to ensure its creation becomes essentially carbon neutral, to the following charities, in equal shares for global-warming-related issues:

1 WWF.
2 The Alliance for Climate Protection.
3 Global Cool.
4 Pure, the clean planet trust.

I sincerely hope that this book has scared you, as watching Al Gore's *An Inconvenient Truth* certainly concerned me! More importantly that you make a decision to do something to alleviate the global-warming problem facing all of us on planet Earth.

While anthropogenic (manmade) global warming may well force Earth's temperatures up to potentially life-threatening levels and beyond, if greenhouse gas levels are not stabilised it is perhaps important to remember that planet Earth during its ancient history has faced many challenges and disasters. Earth has survived comet strikes, meteor impacts, sea-level rises and of course ice ages, so it *will* survive. The question is, will we?

Another ice age will no doubt arrive eventually, as it has done repeatedly for millions of years, as discussed in Chapter M. Until that time, the human race must make every conceivable effort to switch energy production away from fossil-based fuels and create the technology to produce zero-emission energy, to avoid the unpleasant predictions of a world that is 2°C or more warmer than in pre-industrial times. We all may have to adapt, to some extent to a warmer world. We must all hope that all that can be done will be done during the ten-year window that is left to prevent what could be the beginning of the end of life as we know it on planet Earth. No doubt the scientists of the future, if there are any left, will not be concerned with global warming, but global cooling instead, as the next glacial cycle takes grip, but that's another story…

Key facts

- ➢ It is thought that there is a window of only ten to fifteen years to make significant cuts in global-warming greenhouse gases, to prevent serious warming of Earth's climate and any potential catastrophic effects that may flow.
- ➢ It is recognised that temperatures should not be permitted to rise by more than 2°C (3.6°F) above pre-industrial levels, to avoid dangerous climatic change from occurring.
- ➢ To prevent this, CO_2 levels must not go above 400 ppm. They are already at 385 ppm, and rising at the rate of up to 1.9 ppm each year, meaning 400 ppm may be reached in just eight years' time.
- ➢ If nothing is done, then by 2050 greenhouse gas levels, which are rising at about 2.5 ppm, will be at about 550 ppm, forcing Earth's temperature up to 2–5°C (3.6–9°F) warmer than pre-industrial times.
- ➢ Such a temperature increase could mean a rising risk of the West Antarctic ice sheet collapsing, a rising risk of a collapse of the ocean thermohaline circulation, and potential for the Greenland ice sheet to start melting irreversibly. If these things happened, the world would know about it…

---

1 WWF, <www.panda.org> (on shifting the energy paradigm).
2 <www.hubbertpeak.com>.
3 <www.virginearth.com>.
4 Stern Review on The Economics of Climate Change, Part II.

Appendix

Some Useful Web Addresses

<www.a-zofglobalwarming.com/>
<www.cafepress.com/globalwarmin> (to purchase various gift items with colour illustrations from this book)
<http://a-zofglobalwarming.blogspot.com/> (for the latest news and information on global warming and climate change)

Notes on Cover Art

Blue Marble East Photograph of the Earth (front cover)

This spectacular 'blue marble' image is the most detailed true-colour image of the entire Earth to date. Using a collection of satellite-based observations, scientists and visualisers stitched together months of observations of the land surface, oceans, sea ice, and clouds into a seamless, true-colour mosaic of every square kilometre (386 square miles) of our planet. (NASA visible earth, <www.visibleearth.nasa.gov>.)

Earthrise (back cover)

This view of the rising Earth greeted the Apollo 8 astronauts as they came from behind the moon after the lunar orbit insertion burn. The photo is displayed here in its original orientation, though it is more commonly viewed with the lunar surface at the bottom of the photo. Earth is about five degrees left of the horizon in the photo. The unnamed surface features on the left are near the eastern limb of the moon as viewed from Earth. The lunar horizon is approximately 780 kilometres from the spacecraft. Height of the photographed area at the lunar horizon is about 175 kilometres.

Astronauts Frank Borman, Jim Lovell and William Anders had become the first humans to leave Earth's orbit, entering lunar orbit on Christmas Eve 1968. In a historic live broadcast that night, the crew took turns reading from the Book of Genesis, closing with a holiday wish from Commander Borman: 'We close with good night, good luck, a merry Christmas, and God bless all of you – all of you on the good Earth.' (NASA, <www.nasa.gov>.)

INDEX